WIRELESS PERSONAL COMMUNICATIONS

IEEE PRESS
445 Hoes Lane, PO Box 1331
Piscataway, NJ 08855-1331

Karen Henley, *Production Editor*

Page composition by
Lee James Communications

Cover design by
Peter Jeziorski

Wireless Personal Communications

The Future of Talk

Ron Schneiderman

The Institute of Electrical and Electronics Engineers, Inc., New York

This book may be purchased at a discount from the publisher when ordered in bulk quantities. For more information contact:

IEEE PRESS Marketing
Attn: Special Sales
PO Box 1331
445 Hoes Lane
Piscataway, NJ 08855-1331
Fax: (908) 981-8062

Printed in the United States of America

10 9 8 7 6 5 4 3 2 1

ISBN 0-7803-1010-1

IEEE Order Number: PC0359-0

Library of Congress Cataloging-in-Publication Data

Schneiderman, Ron.
Wireless personal communications : the future of talk / Ron Schneiderman.
p. cm.
Includes bibliographical references and index.
ISBN 0-7803-1010-1
1. Wireless communication systems—United States—Technological innovations. 2. Mobile communication systems—United States—Technological innovations. I. Title.
HE8664.S36 1994
384—dc20 94-4438
CIP

For Susan, who wanted to edit this book;
Dr. Todd, who wanted to dissect it,
and Josh, who wants to make a movie out of it.

CONTENTS

Preface

People love to talk, and at the rate things are going they have never had a better opportunity. There are more cellular and cordless telephones currently in use around the world than ever before and the market is growing fast (the cellular market is growing at the rate of almost 50 percent a year). At the same time, new personal wireless communications products and services are being introduced around the world.

As this book goes to press, the Personal Communications Industry Association was projecting a total of 88.3 million subscribers to New Personal Communications Services (PCS), Paging, Cellular, Enhanced Specialized Mobile Radio (ESMR/SMR), Dedicated Data, and Mobile Satellite Services by 1998—climbing rapidly to 167.4 million subscribers in 2003. That's a huge jump over the 33.7 million subscribers to wireless services in 1993.

Data, barely 1 percent of the wireless market today, is poised to explode. Wireless personal communicators, which are expected to account for a major segment of the data market, may never become quite as ubiquitous as the wrist watch or even the calculator, but their utility in a world where information is life itself could very well create the fastest growing consumer electronic market since the introduction of the transistor radio.

The market offers a huge new opportunity for the global electronics industry, including telecommunications, semiconductor/components, test and measurement, and computer equipment manufacturers; network service providers, software developers, distributors, and retailers.

This book offers a very detailed look at wireless communications. It covers major issues, such as market projections, strategies, and alliances; product development programs and patterns; emerging technologies; regulatory issues; software developments, and standards, as well as long-term opportunities in personal wireless communications.

As one trade magazine reporting on the current boom in personal wireless communications put it, “If it moves, you may be able to communicate with it.”

It is dangerous to put limits on wireless.

Guglielmo Marconi (1932)

Chapter 1

An Overview of Markets, Products, and Services

There's a wonderful scene in the Woody Allen film *Play It Again, Sam* in which actor Tony Roberts, who plays a hardworking and somewhat obsessed businessman, calls his office frequently to make sure everyone knows where he is at all times. "Let me tell you where you can reach me, George. I'll be at 362-9296 for awhile. Then I'll be at 648-0024 for about 15 minutes. Then I'll be at 752-0420, and then I'll be home at 621-4598."

At which point, co-star Diane Keaton, looking exasperated at the Roberts character, says, "There's a phone booth on the corner. You want me to run downstairs and get the number? You'll be passing it."

A very funny bit, but also very prophetic. In the new world of personal telecommunications, the wire in the wall will no longer dictate where we must be in order to use a telephone. Calls will go to people, not places. The caller will not even have to know the location of the person being called. Everyone will need only one phone number. You can be reached by phone by anyone, anywhere, at anytime.

The generic terminology for this fast-emerging market is "wireless personal communications," it's the new hot button in telecommunications and consumer electronics.

"Wireless" has all the elements of a fantasy market: tremendous growth potential, global in scope, technology-driven, and capable of producing a virtual endless stream of new products to a market that cuts across commercial, consumer, even military applications. The Cellular Telecommunications Industry Association (CTIA) believes that over the next decade, "wireless devices and wireless networks will be the dominant mode of communications in the United States." Arthur D. Little, Inc., agrees. The Boston-based management and technology consultant has told its clients that half of all telecommunications traffic in North America will be wireless by the year 2000.

In more down-to-earth terms, the cellular industry's vision of the future looks something like this:

- A corporate executive in a hotel room hundreds of miles from home puts the finishing touches on her proposal for a new product line. She hits the "transmit" key on her notebook computer. No wires, no modems. The proposal zips through the air back to the home office.
- A student is sitting in the school library doing her homework. Her electronic notepad contains a calculator, encyclopedia, dictionary, science assignment, and even her homework. The message light flashes; she presses it. "Don't forget to be home by 4:00. You have a dentist appointment. Love, Mom."
- You're driving down a dark and icy road. You lose control of your car, run off the road, and hit a tree. Your air bag inflates instantly, transmitting an emergency message and your precise location to a nearby ambulance service.

Ironically, independent market researchers and industry analysts, who frequently go overboard in their market projections, have been much too conservative in estimating the rate of growth of wireless personal communication products and services. When cellular technology was inaugurated in the United States just over 10 years ago, McKinsey & Co. told AT&T, which invented cellular, that no more than 900,000 cellular phones would be in use by the year 2000. At the rate the cellular industry is signing up new subscribers, McKinsey's forecast may be off by about 20 million. How could McKinsey be so far off in its projections? For one thing, its projections were based on technologies and market demand assumptions that existed at the time, when a cellular telephone was a 25-pound piece of equipment installed in a car trunk, cost almost $3,000, and service was limited to only a few major market areas.

Even more recent, and presumably more sophisticated, market forecasts have completely missed the mark. Dataquest Inc.'s 1989 forecast of cellular phone users in the U.S. for 1991 was nearly half the actual number for that year when more cellular phones were put into service than wired residential phone lines—2.5 million versus 1.9 million. By the end of 1992, there were more than 11 million cellular subscribers in the U.S. alone. That's enough "early adopters" to put the cellular telephone in the same league as the VCR as the world's fastest-selling consumer electronics product.

The most recent market projections are staggering. Motorola, the world's leading supplier of telecommunication equipment, expects the market for wireless equipment and services to grow to $600 billion by the year 2010; that's 20 percent of the projected $3 trillion worldwide telecommunications market. BIS Strategic Decisions, an international technology consultant and market analyst, believes the number of cellular and cordless phones worldwide will grow from 22.5 million in 1992 to 40.9 million in 1996. At the same time, The Gartner Group says it expects the overall mobile communications satellite market to surpass cellular telephones and digital paging combined, and will ultimately match the installed base of personal computers.

Cellular phones continue to sell at a record pace, with more than 9,000 new subscribers signing up for cellular service in the U.S. every day, bringing the total number of subscribers at the end of 1993 to more than 13 million in the U.S. alone. Paging, with more than 11 million users in the U.S., remains the technology of choice for those who need mobile communications, but do not require voice communications, and it is rapidly penetrating the consumer market with new features and products.

But new wireless services are emerging:

- Mobile satellite services (MSS), which are still in the technical development and regulatory stage, could evolve into a multibillion-dollar business by the turn of the century.
- Specialized mobile radios (SMR), used by fleet dispatchers such as taxis and delivery services and led by companies like Motorola and NEXTEL Communications (formerly Fleet Call), have only

around 1.4 million users, but that's about to change. The Federal Communications Commission (FCC) has decided to allow NEXTEL to compete with cellular carriers by constructing its own regional, digital wireless network. Other SMRs are moving in the same direction. In time, NEXTEL hopes to interconnect its system with other SMR operators to form a national network.

- Mobile computing, only now beginning to emerge for the masses, suffers from high prices and a lack of standards. Once those problems are solved, it's going to be another huge market with tremendous potential for new product development and services.

- Personal Communications Services, or PCS, the highly touted "anyone, anywhere, anytime" service, is an ill-defined concept that will likely consist of a new class of very small telephone handsets that combine some of the best features of cordless, cellular, and basic telephone services, including data communications. PCS phones are expected to be much cheaper than cellular, but with less range than cellular phones. Based on that description, The Personal Communications Industry Association believes that if the FCC licenses PCS by 1994, as expected, the market could reach 56 million subscribers by the year 2002. In Arthur D. Little's scenario, a shirt-pocket communicator that costs less than $50 a month to operate could generate annual revenues of $40 billion within 10 years after market launch.

PCS PROJECTIONS**

	1991	1992	1993	1994	1995	1996	1997	1998
Cellular Subscribers	7.6	10.5	13.5	16.5	20.0	24.0	28.0	32.0
PCS	0.0	0.0	0.1	0.5	1.0	1.5	2.1	2.6
Paging Subscribers	11.5	13.2	15.2	17.2	19.2	22.0	24.0	26.0
Mobile Data	0.25	0.50	1.2	2.1	3.0	4.3	5.7	7.4
Mobile Computers*	0.0	0.0	0.023	0.075	0.255	0.795	1.515	2.595
PDAs	0.0	0.0	0.002	0.022	0.094	0.289	0.889	1.795

Source: The Yankee Group, 1993
* Portable computers with integrated wireless communications capability
** Users in millions

The Yankee Group is projecting enormous growth for all categories of wireless personal communications products.

More and more, these services will begin to overlap and compete. EO, Inc.'s notebook-size pen-based wireless communicator, for example, features a built-in cellular phone. Apple Computer's highly-touted Newton MessagePad was introduced without a communications capability, but that's being fixed. Pricing will become a larger issue as the market shifts from professional and business users to the mass market. This is already evident from the declining national average of monthly bills of cellular subscribers—down from $95 three years ago to $68 today.

PCS grew out of a report published in Britain in 1989 called "Phones on the Move: Personal Communications in the 1990s." Barely seven pages long, the report got the telecommunications community in the United Kingdom to thinking that the U.K. had the potential to become a "world leader" in mobile telecommunications. "More and more," the report stated, "U.K. business is coming to rely on mobile communications, and government has acted as an enabler, making sure they get the services they need...." The British government didn't disappoint. It quickly licensed four companies to provide so-called telepoint services. Also known as CT-2 (cordless telephone—second generation), telepoint is essentially a cordless pay phone, allowing subscribers to originate, but not receive, short-range phone calls in public areas equipped with telepoint base stations, such as shopping centers, train stations, and airports.

Telepoint seemed like a good idea at the time; London has very few public pay phones. But high subscriber costs ($200 for a handset, plus a $60 service connection fee and a monthly service charge of $15) didn't play well in a weak British economy. Also, the systems licensed by the government were not compatible. Today, only one company is offering limited telepoint service in the U.K.

With all its problems, the introduction of telepoint in the U.K. sent a wake-up call to the rest of the world. In the U.S., the FCC responded by issuing more than 200 experimental licenses for PCS trials. Cellular telephone carriers, cable television system operators, and independent telecommunications operators are spending millions of dollars to develop new products and services based on PCS concepts. Europe's cellular phone market hasn't grown as rapidly as the U.S. market, possibly because many countries have different technical standards, making "roaming" between countries virtually impossible. Car phones that work in Germany, for example, won't work in France, and vice versa. That's changing with the introduction of the Global System for Mobile Communications (GSM), the digital cellular standard for Europe established by the European Telecommunications Standards Institute (ETSI) (see Chapter 6).

Europe offers a unique opportunity for new telecom services because there are fewer entrenched players and spectrum use is lighter and less of an issue than in the U.S. Japan's cellular telephone market has grown at the rate of 80 percent annually in each of the last three years. But its market penetration rate remains low—slightly more than 1 percent of Japan's population uses a cellular phone. (At the end of 1992, the total U.S. penetration rate was 4.37 percent, and the average penetration rate in major metropolitan markets was 5.15 percent.) Nippon Telegraph & Telephone (NTT), which accounts for about 550,000 of the country's 870,000 cellular subscribers, expects the market penetration in Japan to double by the year 2000. As a result, virtually every major U.S. cellular system operator and equipment supplier is pursuing these markets.

Based on an optimistic timetable for FCC action on spectrum allocation for PCS and an ambitious rollout schedule for commercial PCS service, the FCC expects to begin granting PCS licenses in early 1994. The questions the industry is wrestling with today are, "Who really needs, or wants, PCS?" "Who will use PCS?" "Can portable phones get much smaller than the 5.9-ounce, cigarette pack–size models that are available today?" "If the service isn't a lot cheaper than cellular and doesn't offer the same features as cellular, won't PCS providers have a tough time shifting cellular users to less functional services?"

There's also a burgeoning market in wireless office equipment, with wireless local-area networks (WLANs) and wireless private branch exchanges (WPBXs) getting most of the attention.

WLANs, which can transfer data and share resources such as printers without physically connecting them, have received most of the attention from equipment manufacturers because they represent two of the fastest-growing segments of the computer industry—local-area networks (LANs) and mobile computing. The big advantage with WLANs is that they eliminate the cost of rewiring an office every time it is reorganized or repartitioned. It's already a fast-growing market: Market research by International Data Corp. (IDC) and studies on worker mobility by Steelcase, a major supplier of office furniture, estimate the number of wireless connections in U.S. offices will grow from 68,000 in 1992 to 538,000 in 1994, and to well over 1 million in 1995. Still, few computer system managers would be willing to rip out their wired LAN, which represents a significant investment and can move data much faster than wireless LANs.

WPBX offers business users the ability to make and receive calls using cordless phones anywhere on or near company premises. As with WLANs, wireless WPBXs should increase the use of computers, including laptops and the even smaller notebook models, and should help satisfy what the North American Telecommunications Association (NATA) says is the "enormous pent up demand" for wireless business communications systems.

Needless to say, industry companies are getting the message:

- American Telephone & Telegraph Co., the largest long-distance telephone company in the U.S., has announced plans to acquire McCaw Cellular Communications, the largest cellular carrier in the country, with operations in more than 100 cities.
- Motorola, the largest telecommunications equipment supplier in the world, has organized an international consortium of more than 30 companies to finance and launch its global satellite Iridium network.
- American Mobile Satellite Corp. (AMSC) has signed agreements with 60 cellular-service providers, allowing them to offer their customers access to AMSC's mobile satellite services anywhere in North America.
- MCI Communications Corp., the second largest long-distance carrier in the U.S., has formed a consortium of 200 companies to create a national PCS network. MCI and British Telecommunications Plc (BT) are also working toward the joint development of common international telecommunications products and services, with MCI covering the Western Hemisphere and BT focusing on the Eastern Hemisphere. BT plans to invest $4.3 billion in MCI, giving it a 20 percent stake in the U.S. company, putting MCI in a stronger position to compete with its U.S. long-distance rival, AT&T. MCI has already said that much of the new BT funding

will go to its "multimedia strategy," including wireless applications.

AT&T plans to invest $3.8 billion for a 33 percent stake in McCaw Cellular Communications. Once the negotiations are completed (with government approval), AT&T said it would acquire all of McCaw Cellular for $12.6 billion in stock. The deal would almost immediately link McCaw's 2 million cellular customers with AT&T's 90 million customer long-distance network.

The arrangement also gives McCaw use of the AT&T brand name in marketing wireless services in North America, as well as access to AT&T's marketing, sales, customer service, and distribution channels, and to AT&T Bell Laboratories, one of the most respected R&D facilities in the world. McCaw also owns stakes in Lin Broadcasting Co., a cellular and television broadcast firm; American Mobile Satellite Corp., the first mobile satellite service in the U.S., and Claircom, Inc., a joint venture with Hughes Network Systems, which provides telephone service to commercial and private aircraft. The deal would also require British Telecom, which owns 22 percent of McCaw, to sell its interest to AT&T.

Of course, AT&T already sells cellular equipment and produces electronic components for wireless communications applications. And it has strategic R&D, distribution, and manufacturing alliances with Oki Telecom, the U.S. arm of one of Japan's biggest telecommunications equipment producers; Matsushita Electric of Japan, a major cellular phone and components manufacturer; General Magic, an Apple Computer spin-off set up to develop highly innovative multimedia wireless products; and EO, Inc., a California start-up that is already shipping a new class of mobile, pen-based (cellular optional) personal communications devices. AT&T has also taken a majority investment position in Go Corp., a leading producer of pen computing software, which has become a subsidiary of EO.

The agreement could speed the development of a truly seamless, nationwide wireless communications network. By linking up with McCaw, AT&T can avoid routing its calls through local telephone companies, adding to its own profitability while cutting off a critical revenue stream of the regional telcos. That would put AT&T in direct competition with the regional Bell telephone operating companies (RBOCs) for the first time since the breakup of the Bell System in 1984. The cost of the AT&T/McCaw connection to the RBOCs could be enormous: Access charges paid to local telephone companies are about 40 percent of the cost of providing long-distance service and account for an estimated one quarter of the $100 billion local phone industry.

If long-distance companies can compete for local access, can local phone companies compete for long-distance business? Clearly, they're trying. Three of the RBOCs, sometimes known as the "Baby Bells," have asked the FCC to investigate AT&T's proposed alliance with McCaw Cellular. In a joint petition to the FCC, NYNEX, Southwestern Bell, and Bell Atlantic claim that the bid by AT&T to buy into McCaw would put AT&T back into the local telephone business, which they view as a violation of the 1984 consent decree that broke up AT&T. However, rather than proposing that the FCC block the AT&T/McCaw transaction, the Baby Bells asked the commission to repeal restrictions on their own business activities or apply similar restrictions to AT&T and McCaw.

AT&T's position is that its acquisition of McCaw would not change the local exchange monopoly held by the seven regional Bell operating companies. McCaw's response to the petition was that it was "self-serving" and that the AT&T/McCaw merger is "fully supportive of public telecommunications policies and will increase competition in the wireless communications industry, bringing new services and infrastructure capabilities to the American public."

The RBOCs have also begun to reposition themselves by forming wireless groups to focus on emerging new wireless services. San Francisco–based Pacific Telesis Group has won approval to split off its cellular operation into a new business. It will be barely one-tenth its original size, but it will also be out from under the highly restrictive 1984 accord that broke up AT&T and created PacTel as a regulated local telephone exchange service. The move also puts the new organization, named PacTel Wireless, in a position to form joint ventures and even produce its own equipment, and it creates more of a "pure play" for investors interested in the fast-growing wireless market.

MCI has teamed up with Bell Atlantic Mobile Co. to test a "Follow Me 800" service that delivers calls to cellular customers in their cars virtually anywhere, and it is organizing a national network of companies to build and operate a PCS network that would allow its customers to make and receive calls anywhere in the country. To do that, MCI has asked the FCC to license three national consortia rather than individual companies. Each consortium would have a national manager and local operators. The manager would provide network services, technical standards, marketing, national roaming capabilities, and interoperability among the systems. MCI also wants the FCC to classify PCS operators as common carriers with the status of co-carriers with local telephone companies. This would entitle PCS operators to the same kinds of interconnection and traffic exchange agreements that exist among local telephone companies. MCI would hold a minority interest, but would be the national manager of the PCS network; the majority interest in the consortium would be held by companies providing local personal communication services.

So far, MCI has put together a broad-based consortium of almost 200 companies to provide PCS in areas serving half the U.S. population, including the contiguous U.S., Alaska, and Puerto Rico. If the FCC allows the participation by consortia in the licensing of PCS and if national PCS licenses are granted, the consortium will file for a national PCS license with the FCC.

It's a diverse group that includes cable companies, independent telephone companies (ITCs), paging operators, small wireless experimenters, and others. The list, which continues to grow, includes:

ALLTEL
Barden Communications
Columbia International
Concord Telephone
Crico Communications Corp.
GCI
Iowa Network Services (131 ITC companies)
Jones Lightwave
Kansas Personal Communications, Inc. (21 ITC companies)
Landmark Communications
Lexington Telephone
Minnesota Equal Access Network Services (58 ITC company shareholders)
North State Telephone
Northwest Iowa Telephone
Prairie Grove Telephone
StarPage
Susquehanna Cable Co.
Systems Engineering Management Associates, Inc.

Teleport Denver
Times Mirror Cable Television
TPI Communications International
TRI Touch America (a subsidiary of Montana Power)
Wisconsin Wireless

MCI also has a shot at competing with New Jersey Bell for local telephone service. In early February 1993, the state's Supreme Court overturned a 1990 ruling by New Jersey regulators who had rejected MCI's application to offer local service in addition to toll and long-distance calls. MCI's "friends and family" promotion, which provides discounts to people who call specific numbers, could be extended to calls in the local area if MCI wins approval from state regulators. A similar application filed in 1992 by the Sprint Corp., the number three long-distance carrier in the U.S., is pending. So far, 40 states already allow competition for local calls. Others are considering it.

Sprint, meanwhile, has its own wireless strategy, acquiring Centel Corp., which owns substantial cellular properties as well as local phone companies.

Ironically, cable television system operators, which fought for years to keep telephone operating companies out of their business, have become increasingly involved in cellular and hope to carve out a large niche for themselves in PCS. At least 30 cable companies have already obtained licenses from the FCC to build and test wireless phone networks using cable wiring and cellular phones to produce advanced, but presumably low-cost, PCS systems. One of them, Cablecast Corp., is conducting a two-year, five-city PCS experiment to determine whether its cable systems can carry telephone calls linked with cellular phone systems. On an even grander scale, Cable Television Laboratories, Inc., an R&D consortium that represents 85 percent of the cable TV system operators in the U.S., is working with AT&T Bell Laboratories and Arthur D. Little in a program to integrate PCS into cable television systems.

One of the reasons for the fast growth of the U.S. cellular phone market is that, unlike Europe, the U.S. has a single standard—the Advanced Mobile Phone System (AMPS)—an analog system operating at 824–849 megahertz (MHz) and 869–894 MHz. But the analog technology has limited capacity compared with emerging digital systems.

Europe and Japan have already moved ahead of the U.S. in the implementation of advanced digital cellular telecommunications. The European Community has settled on the GSM as its digital cellular system, and the Digital European Cordless Telecommunications (DECT) system as its digital cordless standard, and it is developing several personal communications network services.

Clearly, there will be intense global competition in Europe, Japan, and North America over the next several years to establish entirely new standards for wireless communication systems. In the U.S., the FCC governs U.S. spectrum allocations and use, but the Telecommunications Industry Association (TIA) develops and specifies most telecom standards. In Europe, the ETSI is responsible for approving the emerging Pan-European standards for the European Community. Japan's closest equivalent to an FCC or TIA is the Ministry of Post and Telegraph (MPT). However, most of the important work in arbitrating the rapid movement toward new telecommunications services was done at the 1992 World Administrative Radio Conference (WARC '92) in Spain.

WARC established new frequency spectrum allocations for a wide range of satellite communications services, and its decisions will present a variety of challenges in the design and definition of future satellite communications services and systems. Key in WARC's efforts were allocations of frequency bands to the Mobile Satellite Services (MSS) in the 1- to 3-gigahertz (GHz) range, including low-earth-orbit (LEO) MSS systems and Future Public Land-Mobile Telecommunications Systems (FPLMTS), an international microcellular "standard" planned for a post-2000 time frame. WARC officials have already "reserved" the frequency bands 1885–2025 MHz and 2120–2200 MHz for FPLMTS, which overlap portions of the new MSS spectrum allocations.

Despite advances in the technology and a better understanding of the forces that are likely to drive the market, questions remain for both wireless personal communications equipment manufacturers and service providers. For example, what are the primary dynamics of the market? What new distribution strategies will be necessary to become a full participant in the market? How can lessons learned from current data-over-cellular and PCS trials be applied to product development and marketing strategies? How can cellular and PCS co-exist in the marketplace? What role will the cable television industry play in the wireless personal communications market? How will top-line telecommunications equipment manufacturers respond to the demand for consumer-level products and pricing? How will the industry deal with market changes and technologies that outpace the regulatory structure? And the most important questions of all: What types of products and services do people really want, and what are they willing to pay?

Also, where does wireless communications fit into the nation's much-discussed information superhighway? Two congressmen, George E. Brown, Jr. of California, chairman of the House Science, Space, and Technology Committee, and Rick Boucher of Virginia, chairman of Brown's Subcommittee on Science, have asked the Congressional Office of Technology Assessment (OTA) to study the role that wireless communication technology might play in the national information infrastructure.

"If we're not careful," says Boucher, "wireless technologies could well be the 'missing link' in the construction of a national information infrastructure that is capable of fully servicing all Americans."

It may be easier to answer most of these questions in the short term; longer range, things get a little muddy. It will be very difficult to forecast technological advancements and how consumers will respond to the myriad choices of products and services that hit retailers' shelves. Nicholas Negroponte may be right. The controversial head of MIT's Media Lab predicts that what is now wired will eventually become wireless and virtually everything now transmitted over the airwaves, or wirelessly, will be wired, including television and telephones. It's called the Negroponte Flip. Given the high stakes (the AT&T and McCaw deal, for example) and the intensity of the competition—even at this nascent stage in the development of the market—there isn't going to be much time to figure it out.

CHAPTER 2

CELLULAR-SIZZLING SALES, HOT NEW PRODUCTS

If you looked closely at the sketch of the futuristic cellular telephone in the long-running Sharp Electronics advertisement, you could see a hand-held unit with a keyboard to operate a telephone, personal computer, and calculator; buttons to control fax and electronic- and voice-mail functions; two postage stamp-size video screens (one to see the person you're talking to, the other to display graphics); tiny knobs to handle audio and video features; and a built-in, miniature CD-ROM drive. "It's the cosmic communicator," says the ad, "but it's not pie in the sky." Not any more.

Today, you can use your cellular phone to call just about anywhere in the world. In some places you can even transmit data over your cellular phone. Before long, you will be able literally to call anyone, anywhere at anytime—on your cellular phone. Eventually, you will be able to perform most of the functions depicted in the Sharp ad.

The steady stream of smaller, cheaper, more functional cellular phones has built the cellular industry from scratch just 10 years ago to 13 million subscribers at the end of 1993. The market is growing at the rate of 40 percent a year. The Cellular Telecommunications Industry Association (CTIA) estimates that half

"It's table 44 on his cellular phone. He wants a waiter."

From *The Wall Street Journal*—permission, Cartoon Feature Syndicate.

of all new phone "installations" in the U.S. between 1995 and 2000 will be cellular. Projections by EMCI, Inc., a market research and consulting group, indicates there will be 28 million cellular phone subscribers in the U.S. by the end of 1997. The average penetration of the top 17 cellular carriers in the U.S., which EMCI says account for nearly 90 percent of all U.S. cellular subscribers, was 2.48 percent in 1992, compared to the previous year's penetration of 1.97 percent.

Why is it called "cellular?" In nontechnical terms, cellular works by dividing a city or region into small geographic areas called cells, each served by its own set of low-power radio transmitters and receivers. Once a cellular call or data message reaches a transmitter/receiver tower, it is plugged into the regular land-line phone system. Each cell has multiple channels to provide service to many callers at one time. As a caller moves across town, the signal to or from the cellular telephone is automatically passed from one cell to the next, without interruption.

AT&T Bell Laboratories developed the concept in 1947, but it wasn't until 1962 that the first tests were conducted to explore commercial applications. It then took another eight years before the Federal Communications Commission (FCC) set aside new radio frequencies for "land mobile communications." That same year (1970), AT&T proposed to build the first high-capacity cellular telephone system. It was called the Advanced Mobile Phone Service, or AMPS.

The FCC decided to license cellular systems in the 306 largest metro areas first (called metropolitan service areas, or MSAs), then to the less populated 428 rural service areas (RSAs). With its rules in place, the FCC began accepting applications for the 60 largest cities during 1982. In early 1983, when 567 applications were filed just for the 30 markets ranked 6lst to 90th in size, the FCC knew it had a problem; its traditional system of issuing licenses following comparative hearings would take forever. So in May 1984, the commission amended its rules to allow lotteries to be used to select among competing applicants in all but the top 30 markets. On October 13, 1983, the first cellular system began operating in Chicago.

By mid-1993 there were 11,551 cell sites, supporting 1,506 cellular systems in the U.S. Each MSA and RSA is served by two carriers: a wireline (or B-side) company, such as a Bell operating telephone company, and an independent, nonwireline (A-side) company. In organizing cellular in this manner, the FCC was trying to create competition between telephone companies and newcomers to the industry. In 1992, with the inauguration of cellular service in rural northeast Mississippi, the cellular industry reached every market in the U.S., extending into all 734 markets across the nation. (Internationally, 67 countries have cellular systems and that number is growing rapidly.)

The following list represents the 25 largest U.S. cellular operators, ranked by the combined population of the markets for which they hold a majority interest in licenses. The population total (POPS in industry terminology) listed for each company reflect the figures assigned to the carrier by Donaldson, Lufkin & Jenrette's Winter 1992–93 edition of "The Cellular Communications Industry."

THE TOP 25 U.S. OPERATORS	POPS (in millions)
1. McCaw Cellular Communications	61.0
2. GTE Mobile Communications	54.2
3. BellSouth Cellular Corp.	39.1
4. Bell Atlantic Mobile	34.7
5. PacTel Cellular	33.0
6. Southwestern Bell Mobile Systems	32.5
7. Ameritech Mobile Communications	21.4
8. NYNEX Mobile Communications	19.5
9. United States Cellular Corp.	19.5
10. US WEST NewVector Group	17.6
11. Sprint Cellular	15.9
12. Cellular Communications, Inc.	7.8
13. ALLTEL Mobile Communications	7.6
14. Comcast Cellular Corp.	7.2
15. Vanguard Cellular Systems	6.0
16. Century Cellunet, Inc.	5.6
17. Centennial Cellular Corp.	3.9
18. Associated Communications	3.6
19. Cellular, Inc.	3.4
20. Puerto Rico Telephone Co.	3.4
21. SNET Cellular	3.3
22. Cellular Communications of Puerto Rico	2.8
23. Cincinnati Bell	2.2
24. Sterling Cellular	2.2
25. Cellular Information Systems	1.8

The figures for Comcast, Puerto Rico Telephone, Sterling Cellular, and Cellular Information Systems are based on CTIA records of information filed with the FCC prior to December 31, 1992, and the association's semiannual *Data Survey* results for year-end 1992.

Most cellular systems in the U.S. and the rest of the world today are analog networks, a technology that uses a continuous electrical signal that varies in frequency and/or amplitude rather than a pulsed or digital signal. More than half of the world's cellular subscribers are using the AMPS system developed by AT&T. As a standard, AMPS has worked very well, providing cellular carriers with significant economies of scale in equipment manufacturing, while allowing users to "roam," or make calls outside their home service areas.

Now, cellular carriers and equipment manufacturers are preparing to upgrade their systems from analog to digital technology. The idea is to offer a higher-quality, higher-capacity, and more feature-rich service. However, it is not yet clear which of a number of proposed digital standards will serve the U.S. cellular market.

The CTIA has spent more than five years trying to rally the U.S. cellular industry behind a technology known as time-division multiple-access (TDMA), which works by dividing time at a channel frequency into parts and assigning different phone con-

Motorola's MICRO TAC portable flip-phone is one of the smallest and lightest cellular phones on the market.

versations to each part. At the outset, TDMA, which claims three times the transmission capacity of present-day analog cellular systems, had virtually total industry support, mainly because it was a fairly well-understood technology. The CTIA even adopted TDMA as its digital transmission standard in 1990 and continues to support the system. But being the "older" technology in today's fast-moving wireless personal communications market has actually slowed its acceptance while other, more advanced, digital cellular systems have won support from cellular carriers and equipment suppliers.

TDMA's strongest competitor is code-division multiple-access (CDMA), a proprietary system developed by QUALCOMM, Inc. CDMA is based on a spread-spectrum technique originally developed for the military to scatter signals across a wide frequency band, making it difficult to intercept or jam communications. Other than its antijam qualities, its most important feature is that it offers at least 10 times the communications capacity of present analog cellular systems. CDMA also has an inherent "soft handoff" capability, which reduces the number of dropped calls when passing from one cell to another by connecting the call to the new cell before disconnecting the outgoing cell.

Another digital transmission scheme—broadband code-division multiple-access (B-CDMA)—is being promoted by InterDigital Communication Corp. (formerly International Mobile Machines Corp.). In late 1992, InterDigital acquired SCS Mobilcom, Inc., and SCS Telecom, Inc. With the acquisition, InterDigital picked up the SCS companies' broadband-CDMA technology and its portfolio of 30 patents, some already granted, some still pending. Essentially, B-CDMA works by overlaying and sharing the spectrum with the existing cellular telephone spectrum (825–894 MHz). Like QUALCOMM's CDMA, InterDigital claims that B-CDMA provides additional capacity to the network and improved voice quality.

The CDMA battle was joined in April 1993 when InterDigital filed a lawsuit against QUALCOMM, Oki Electric Industry Co., and its subsidiary, Oki America Inc., charging them with patent infringement. Oki was included because it had announced plans to make CDMA-based phones. The suit, which focuses on a part of the technology that digitally passes a cellular signal from one calling area to another, came at a time when QUALCOMM was making significant progress in convincing the cellular industry that its proprietary narrowband CDMA technique was superior to TDMA. After years of work on CDMA, QUALCOMM was looking forward to collecting licensing fees from virtually the entire U.S. cellular industry as it began making the transition from analog to digital cellular.

But the suit slowed the formal acceptance of CDMA by the industry's standards-setting body, the Telecommunications Industry Association (TIA). Rather than wait out the court, the TIA decided to publish QUALCOMM's CDMA technology as an interim standard, designated Interim Standard-95, or IS-95. The TIA said that its action would at least allow more rapid updating of the standard as manufacturers and users gain practical experience with CDMA technology. But because of InterDigital's patent infringement charge, the TIA placed a cautionary legend on IS-95. QUALCOMM, meanwhile, has asked the court to rule that it has not infringed on any InterDigital/SCS patents.

Like QUALCOMM, InterDigital submitted its B-CDMA technology to the TIA for consideration as a digital cellular standard, but not until the association indicated it was seriously considering the adoption of QUALCOMM's technology as a North American digital cellular standard. The TIA formed a volunteer working group to study B-CDMA a month before it approved IS-95, but no one from industry attended the group's first meeting, and the TIA couldn't find anyone to serve as chairman of the group, at least not immediately. TIA officials suggested to InterDigital that it needed market support to push B-CDMA as an industry standard. (When QUALCOMM approached the TIA two years earlier with its CDMA proposal, it had the formal support of five major cellular carriers.)

Meanwhile, support for CDMA continued to grow. PacTel, US WEST, QUALCOMM, Motorola, AT&T, and a group of other carriers including NYNEX, Ameritech, Bell Atlantic Mobile, Bell Mobility, and GTE Mobilnet performed CDMA tests on QUALCOMM'S San Diego development system. Interviews conducted by Ameritech following side-by-side CDMA/TDMA field trials in suburban Chicago indicated most users preferred CDMA. Nearly all rated overall quality as excellent, and customers perceived significant improvement over their current analog cellular service.

US WEST NewVector Group announced plans to purchase the first commercial CDMA cellular system from Motorola Nortel

QUALCOMM has introduced a mobile cellular phone based on the company's code-division multiple-access (CDMA) digital cellular technology.

Communications and said it expects to have the system in operation by late 1994. US WEST also said it would purchase at least 36,000 dual-mode (CDMA/AMPS) phones from QUALCOMM with such features as a "Help" menu, memory for up to 103 phone numbers with alphanumeric identification, speed dialing, automatic redialing, scratch pad memory, and password with credit card memory number storage. The phone also has a built-in pager. PacTel Cellular, NYNEX, AT&T, Motorola, Oki Telecom, Nokia Mobile Phones, Ameritech Mobile Communications, and Clarion Sales Corp. of America also plan to produce CDMA-based cellular phones under license to QUALCOMM. PacTel will purchase more than 30,000 dual-mode cellular phones from QUALCOMM licensee Oki Telecom to supply its new commercial digital cellular service in Los Angeles beginning in early 1995.

CDMA appeared to win another round when Motorola's Cellular Infrastructure Group, a licensed CDMA equipment manufacturer, announced its next-generation cellular base station in March 1993. The new equipment supports all the current major technologies—Advanced Mobile Phone Service (AMPS), Narrowband AMPS (NAMPS), TDMA, CDMA, and the Global System for Mobile Communications (GSM), the European digital cellular standard. But in its initial implementation for cellular

Hughes Network Systems has developed this dual-mode (analog/digital) cellular telephone based on its unique enhanced time-division multiple-access (E-TDMA) transmission scheme.

carriers in North and South America, Motorola said its new unit would support CDMA and would go to US WEST for its Seattle area system.

CDMA also began to attract attention in Europe, where QUALCOMM has been promoting the technology as a possible alternative to GSM, the TDMA-based Pan-European digital cellular standard. Ericsson, a long-time TDMA proponent, and the largest cellular equipment manufacturer in Europe, has been conducting CDMA field tests for some time and simply needs formal approval from the European Telecommunications Standards Institute (ETSI) to begin producing CDMA-based hardware for European markets. CDMA has already been tested in Switzerland, and three cellular carriers in Australia are studying CDMA in anticipation of heavy cellular traffic in that country's largest cities. Korean telecommunications equipment manufacturers have also shown an interest in producing and exporting CDMA-based portable cellular phones under license to QUALCOMM.

Hughes Network Systems has developed and introduced another digital cellular standard candidate—Enhanced (E-TDMA). Although fully compatible with TDMA, E-TDMA employs digital half-rate voice coding and a digital-speech interpolation (DSI) technique to increase subscriber channel capacity. DSI takes advantage of the quiet times that naturally occur in speech and, in real time, assigns the active speech from conversations to an inactive channel, doubling the call-handling capacity of each channel. According to Hughes, the combination of half-rate voice coding and DSI results in 15 times the analog capacity when trunking efficiencies are taken into account. Hughes claims E-TDMA improves transmission quality over analog because with DSI, a person's speech "hops" between different radio paths, actually reducing the effects of interference or fading. Hughes has also developed and is producing its own E-TDMA handset.

To accelerate E-TDMA's adoption as another industry digital cellular standard, some 200 BellSouth Corp. cellular subscribers in Mobile, Alabama, have been testing E-TDMA–based cellular phones supplied by Hughes. (BellSouth has already launched a commercial TDMA-based service in Los Angeles.) Hughes already has contracts valued at more than $100 million for E-TDMA system installations, including one for $42 million for Tatarstan, a Russian republic, and another from the People's Republic of China.

Some cellular carriers, like Bell Atlantic Mobile Systems, plan to offer customers a digital choice—TDMA or CDMA—even though few people know the difference. McCaw Cellular, the largest U.S. cellular carrier, has committed to TDMA because "it is here today, and it works" and plans to operate a digital network simultaneously with its current analog system in West Palm Beach, Florida, under the Cellular One brand name it shares with other cellular carriers. Customers can convert their phones to the new system or remain on the existing system for a period of time. Southwestern Bell Mobile Systems is another TDMA adherent with a system up and operating in Chicago. Ameritech isn't very high on TDMA, but has been slow to support CDMA.

At last count, at least 11 major cellular phone vendors had either shipped or announced they would ship mobile or portable TDMA-based phones. They include Motorola, Ericsson GE Mobile Communications, Hughes Network Systems, Blaupunkt, Audiovox, Muratec, Nokia, Oki America, Mitsubishi Electronics America, NovAtel Communications, and NEC America.

To get consumers to buy a TDMA/analog cellular phone and subscribe to its digital service, McCaw Cellular has priced the service 20 percent below analog. But manufacturers—after many months of technical and market trials, conferences, and studying reams of independent market research—still aren't sure how the new digital technology will fly in the market. That's clear from interviews with six cellular phone manufacturers by Herschel Shosteck Associates, Inc., a telecommunications industry economist and consultant, who found an enormous divergence in expectations in the use and shipments of digital cellular phones. For 1993, manufacturers were projecting from as few as 100,000 digital telephone sales to as many as 340,000. Says Shosteck, "With expectations so diverse, there is no doubt that some industry participants are positioning themselves for major losses due either to premature commitment or, conversely, to excessive caution."

At least part of the problem is attributable to the standards debate. If you're using a dual-mode (TDMA/analog) phone in Seattle, how can you call someone in California who has a CDMA/analog phone? You can't at the moment, unless the call is automatically switched over to the analog mode. Which explains why cellular carriers expect to offer dual-mode phones for several years.

The interoperability issue becomes more significant as the cellular industry prepares to allow its subscribers to make phone calls anywhere in the country "seamlessly," that is, without first making special arrangements with your carrier. Currently, cellular phones can now place calls virtually anywhere in the United States through "roaming" agreements. But you have to know where you're calling and it can be costly: There is an access charge to roam, which can run from $2 to $3 a day, and an airtime charge—a per-minute charge for each minute, or fraction of a minute—for your cellular call, which can range from 45 cents to $2 per minute.

Under the new system, when a cellular customer travels or "roams" away from home, he or she turns on the phone, the out-of-town system receives the phone's data signal, recognizes that it's not local, and uses the existing Signaling System 7 (SS7) voice-messaging network to talk back to the home system. SS7 is a network switching protocol capable of making high-speed connections. It can also identify the caller and handle remote database interactions, an important feature when incorporated into the IS-41 cellular industry standard for sending messages transparently and seamlessly between different cellular providers. The out-of-town SS7 verifies the customer's identity; checks options like voice mail, call waiting, and preferred long-distance carriers; and then advises the home system to forward all calls.

McCaw Cellular, the CTIA, AT&T, and an organization of 15 U.S. and Canadian cellular carriers have been leading the charge to develop seamless national networks.

McCaw Cellular calls its system the North American Cellular Network (NACN). It picks up where current cellular networks leave off by linking the carrier's regional cellular systems. McCaw started the network in 1991 with the interconnection of the four corners of the country (New York, Florida, the Pacific Northwest, and California). The NACN eliminates access codes, allowing cellular users to be reached anywhere within the network by dialing their regular cellular number. In addition to being able to receive calls, the network offers several features, such as voice mail, call forwarding, call waiting, and conference calling. By late 1993, 35 cellular carriers were participating in the network, reaching into California, Texas, Louisiana, Arkansas, Oklahoma, Kansas, Colorado, Utah, Pennsylvania, New Jersey, and Delaware, increasing the number of customers served by the NACN to 3.5 million across the nation. Cantel's Canadian Network is also hooked into the NACN, which it calls Call Following.

The CTIA went about the process of developing a national seamless system by surveying several technical proposals and then selecting Independent Telecommunications Network, Inc. (ITN), to provide a new national backbone signaling network for its member carriers. Expected to begin operation by late 1993, the service will use ITN's SS7 network to route local and long-distance calls. AT&T's cellular signaling network has been in operation since May 1993 and has been successfully tested by both wireline and nonwireline carriers in several major markets. AT&T said it would provide its SS7-based cellular signaling network to NYNEX Mobile Communications, SNET Cellular, and United States Cellular for a four-month market trial.

A similar service began to take shape when 15 wireline, or B-side, cellular companies, including six of the seven Baby Bells, formed a national cellular network called MobiLink. As is the case with the McCaw, ITN, and AT&T cellular signaling networks, MobiLink users do not require special codes or clearances to place calls while outside their home service area. The system uses standardized dialing and improved calling features through the implementation of the cellular carriers' IS-41 technology.

Other than the obvious advantages of eventually having a high-speed data transmission link between every cellular telephone switching center in the nation and sharing some costs (advertising and administrative costs, for example), the MobiLink consortium believes that having a nationally recognized brand name will be an important factor in promoting their network. Research by McKinsey & Co.—which MobiLink's members sponsored—suggests that a branded national service such as MobiLink could accelerate new cellular subscriptions from among people who don't currently subscribe to a cellular service, but believe that it has some value. According to McKinsey, this is consistent with the pattern for many other telecom products where brand identity plays a larger role in the purchase decision as a market evolves.

The MobiLink consortium includes ALLTEL Mobile Communications; Ameritech Mobile Communications; Bell Atlantic Mobile; BellSouth Mobility; Cellular, Inc.; Century Cellunet; GTE Mobilnet; Contel Cellular; NYNEX Mobile Communications; PacTel Cellular; Rochester Tel Mobile Communications; SNET Cellular; Sprint Cellular; and US WEST Cellular. MobiLink will be offered in Canada by Mobility Canada, which operates Canada's largest wireless network with 12 cellular service providers.

Despite the appearance of competitive conflicts, one MobiLink member, US WEST, owns cellular systems outside its home territory and will compete against the MobiLink service in these areas. Pacific Telesis Group, PacTel Cellular's parent, owns cellular systems with McCaw, which they operate under Cellular One, another national brand. Southwestern Bell Corp. is the Baby Bell that did not join MobiLink, presumably because it is a Cellular One member and uses the Cellular One name outside its home service area.

Motorola has also submitted a proposal to the TIA for a new standard for cellular networks throughout North and South America. Called the A+ Interface, it would allow cellular carriers to install systems using competing vendors' equipment. In the past, the cellular industry had to purchase its cell site and switching hardware from one manufacturer to ensure its interoperability. Motorola says the A+ Interface is similar to the A Interface deployed by a number of leading switch and cell site vendors. The proposal specifies how cellular carriers can manage digital and analog voice traffic between cell site equipment and any switch. Standard interfaces have been used in land-line networks to accelerate the introduction of new services. The A+ Interface presumably would do the same thing for cellular applications.

Typical of how equipment manufacturers get standards accepted by the TIA, Motorola has taken its proposal to the association with some impressive industry support, including endorsements by Northern Telecom, DSC Communications Corp., and Tandem Computer Inc. on the hardware side and NYNEX, BellSouth Cellular, and Ameritech on the carrier side. PacTel Cellular, US WEST NewVector, Sprint Cellular, and Bell Atlantic Mobile have all agreed to test systems that install the A+ Interface.

While most cellular activity has been voice service, there is tremendous interest in using cellular phones and related devices for transmitting data. The reason: Less than 5 percent of U.S. subscribers use their cellular phones for data service, and industry analysts believe that, by 1997, cellular companies may generate as much as 30 percent of their revenue from wireless data transfer. It's going to be an extremely large and competitive market, demanding a constant flow of new products and services.

Cellular Data, Inc. (CDI) thought it had a jump on the industry in early 1991 when it announced that GTE Mobilnet planned to field test its patented cellular packet data system in Houston. The CDI system transmitted packets, or "bursts," of data using the lightly modulated spaces between voice channels to insert narrowband (3-kilohertz), low-power data channels without compromising the quality or availability of voice services. By all indications, the GTE/Cellular Data trial was a success, but CDI could not outrun a group of cellular carriers who got together to develop and promote something called cellular digital packet data (CDPD).

CDPD is an open specification for transmitting packets of data over existing cellular networks at a transmission rate of 19,200 bits per second—up to four times faster than competing wireless services. Unlike the Mobitex wireless data systems used by so-called wide-area networks such as RAM Mobile Data and ARDIS, which use dedicated frequencies outside of the cellular telephone spectrum to serve commercial applications (see Chapter 7), CDPD evolved out of an IBM-developed technology called IBM CelluPlan

II. Simply put, the IBM technique allows packets of data to be sent along idle cellular channels, with the system automatically switching to the least used channel. The complete specification, Release 1.0, was published in July 1993.

The CDPD carriers, which include Ameritech Cellular, Bell Atlantic Mobile Systems, GTE Mobilnet/Contel Cellular, McCaw Cellular, NYNEX Mobile Communications, PacTel Cellular, and Southwestern Bell Mobile Systems, have already established formal relationships with equipment vendors that are expected to propel the deployment of the CDPD network and related mobile computing devices into a dominant market position. McCaw Cellular's Wireless Data Division plans to rapidly deploy CDPD network technology throughout its 105 markets starting in Las Vegas. Several suppliers, including IBM, Apple Computer, Cincinnati Microwave, and EO, have already demonstrated CDPD-enabled products. Network manufacturers with CDPD projects under development include AT&T Network Systems, Cascade Communications Corp., Hughes Network Systems, Motorola, Pacific Communication Sciences, and Steinbrecher.

McCaw's CDPD network, called AirData, will follow with service in New York, Dallas, Miami, Seattle, and San Francisco. Beyond these cities, deployment of AirData will be based on customer demand, until all 105 markets are served, which is expected by the end of 1994.

American Airlines' SABRE Technology Group and the Insurance Value Added Network Service (IVANS), a not-for-profit electronic communications organization, are McCaw's first AirData pilot program clients in Las Vegas. American Airlines plans to demonstrate a wireless SABRE reservation terminal connected over a CDPD link using software developed by the SABRE group. Initially, IVANS plans to test AirData in the Las Vegas area using CDPD-equipped IBM laptop computers. IVANS hopes to expand its reach to major metropolitan markets.

PacTel Cellular's Wireless Data Division, which began a field trial of CDPD service in August 1993, plans to deploy a CDPD network in the San Francisco Bay area. Initially, PacTel's CDPD service will be offered to applications developers and equipment manufacturers to help them speed development and testing of CDPD products. PacTel will start full commercial CDPD service in the Bay area in 1994 and expand it to most of its other markets by the end of the year.

Meanwhile, CDPD products are being introduced at a fairly rapid clip. Cincinnati Microwave, Inc., for example, has teamed with McCaw Cellular to market CDPD products and services to corporate users, and has introduced a wireless CDPD modem aimed at the transportation, utilities, and telemetry industries. Motorola has incorporated a CDPD modem in a PCMCIA card, which transmits at 19.2 kbits/second, the maximum CDPD data rate. Pacific Communication Sciences, Inc. has introduced a CDPD modem designed for IBM-compatible mobile computers.

Motorola is also planning CDPD subscriber products for both consumer and industrial applications. Software companies Alcatel and Retix, a manufacturer of hardware equipment as well, are developing CDPD platforms. Ameritech and Hughes Network Systems are running their own customer trial of CDPD in Chicago, and Ameritech expects to begin commercial CDPD service in its cellular service region in 1994. Meanwhile, Cellular Data has dropped plans to develop its own packet data system and has totally refocused its business; it is now a software company. Its first product is a wireless data network emulator for CDPD networks.

AT&T has licensed yet another technology for transmitting data over cellular voice channels developed by Spectrum Information Technologies, Inc. AT&T and its subsidiaries, NCR and Paradyne, and its new business affiliates, EO and McCaw Cellular, will use Spectrum's technology to develop new families of wireless products. NCR manufactures the wireless-capable Safari laptop computer. Paradyne produces radio modems for cellular phones. EO makes and markets a cellular-based personal communicator. Under the initial agreement, AT&T will pay an initial licensing fee to Spectrum and royalties on unit sales. AT&T also made a $10 million interest-free loan to Spectrum and, under several options developed by the two companies, AT&T may invest $10 million in Spectrum.

Despite all the activity and excitement surrounding digital, analog isn't dead yet. Motorola has been testing its Narrow Advanced Mobile Phone Service (NAMPS), an analog cellular transmission protocol with three times the capacity advantage over current analog cellular systems. NAMPS works by reducing the amount of bandwidth required by each channel from 30 to 10 kHz. The system has already been successfully demonstrated in Japan's NRACS system, which uses 12.5-kHz frequency spacings. Motorola is also working with Texas-based DSC Communications Corp. to develop, with recognized standards organizations, open cellular system interface standards for the international cellular market. By taking an open architecture approach, cellular system operators will be able to purchase cellular system components, such as switches and cells, separately from different vendors.

So far, the cellular industry seems to have everything going for it, even with the introduction of PCS, in which cellular carriers expect to play a major role. What the cellular carriers had not planned on until fairly recently was having to compete with Specialized Mobile Radio (SMR), a private business service using mobile radiotelephones and base stations that communicate through the public phone network. Like cellular, SMR is a two-way service, but it was designed as a private network for communication between a host system and a large number of users, mainly for dispatching taxis and delivery vehicles and public safety.

In February 1991, the FCC gave Fleet Call, Inc. (now NEXTEL Communications, Inc.), the largest independent operator of SMR systems with about 145,000 users, permission to convert its conventional 800-megahertz (MHz) SMR systems in six major markets to digital, cellularlike networks with at least 15 times their current capacity, capable of handling both voice and data. Understandably, the cellular industry, led by the CTIA, McCaw Cellular, and the regional Bell cellular operators, strongly opposed the FCC's decision, fearing that a new service so closely resembling theirs may force them to cut prices. The FCC took a different view, noting that competition is good and that, at any rate, NEXTEL wouldn't be competitive with cellular for years, especially on a national scale. Indeed, SMRs will have to build a totally new infrastructure to handle the new service, including new subscriber handsets, which should make it more expensive than cellular. NEXTEL says its phones will be priced competitively with digital cellular handsets, although the handsets may be a little bulkier than cellular models.

NEXTEL's new system, called Enhanced SMR (ESMR), was launched in the greater Los Angeles metro area in August 1993. It's a huge "footprint," stretching from Santa Barbara to San Diego. Three cellular MSAs cover the same area. The company plans to follow quickly with a system in San Francisco, with service ranging from the northern edge of the Los Angeles system to very close to the Oregon border. Access to NEXTEL is expected in most of California by January 1994. The New York and Chicago metro areas are scheduled to be on-line in mid-1994, with Dallas and Houston activation scheduled for mid-1995.

To cover these markets, NEXTEL must build 400 base station sites. But more important, NEXTEL will have to convince customers to invest in new telephones to use the network. Motorola is the prime contractor and will supply the base stations. Northern Telecom will supply the switches, and Matsushita and Motorola will produce the subscriber handsets.

Financing the system hasn't been easy. NEXTEL's initial capital came from $112.5 million raised in a stock offering in 1992 and funds borrowed from equipment suppliers and financial institutions. It has also agreed to buy a large segment of mobile radio licenses from Motorola for $1.8 billion in stock and announced in November 1993 that Nippon Telegraph & Telephone of Japan would invest $75 million in the company and would design a system that would enable NEXTEL to link all its local systems into a single network within three years. The Motorola licenses, along with those it already owns, would expand NEXTEL's network across 21 states, with the potential to serve 180 million people, making it considerably larger than the McCaw system, currently the largest in the country. However, the NEXTEL ESMR network might not be in operation in any significant way until at least 1996; by that time, there could be more than 20 million cellular users in the U.S., and an almost equal number of pager subscribers.

NEXTEL believes it will need at least $370 million for facilities and equipment through 1995, and much more beyond that date, to expand and operate the network and to cover marketing costs. NEXTEL's prospectus shows that it was operating with a $300 million line of credit from Motorola and Northern Telecom to build its new ESMR system. Additional funding has come from Comcast Corp., a cable television system developer, which said it would buy about $115 million of NEXTEL's stock to maintain its 30 percent stake in the company.

With the recent merger of NEXTEL and Dispatch Communications, Inc., the third largest SMR, NEXTEL's markets now extend from Maine to Virginia in the East and include key markets in the Midwest and Texas, plus most of Arizona. NEXTEL customers will also have access to compatible systems being deployed by Motorola in many of the country's larger markets.

Several other SMRs are right behind NEXTEL. Questar Telecom, Inc. and Advanced Mobilcomm, Inc. have agreed to merge their SMRs in 11 western and southwestern states and offer an ESMR service, including cellular, paging, and two-way dispatch. The merger would create the country's third largest SMR system. Together, the two companies have about 36,000 subscribers. Also, Dial Page, Inc., a specialist in alphanumeric and wide-area regional paging services, also plans to enter the ESMR market. (Motorola operates CoveragePlus, the largest SMR service, which serves mainly long-haul truckers.)

One of the complaints about SMRs has been that, despite their wide geographical coverage, they don't roam. To get around that issue, NEXTEL and several other SMRs participate in the Digital Mobile Network Roaming Consortium (DMNRC), a loosely defined organization that is attempting to build a compatible, national, SS7-based ESMR network, offering services that some cellular networks don't yet offer, such as speed dialing.

EMCI, the market research organization, says the SMR industry has been growing at a rate of about 140,000 new units in service per year since 1985. EMCI expects the growth of traditional analog SMR units to remain flat to slightly lower, but believes that new digital SMR systems will begin to significantly impact the industry beginning in 1994. By 1995, the industry will be adding more digital units per year to the installed base than analog units. EMCI expects Motorola to lead the SMR equipment market and NEXTEL to come out on top as a SMR service provider.

EMCI found that 44 percent of the SMR operators it surveyed toward the end of 1992 already offer mobile data service to their subscribers. Mobile data use among non-SMR private radio subscribers was also believed to be growing, as many private radio users are installing their own private data networks at considerable expense. EMCI estimates that the SMR/private radio technology has a comparative advantage within the dispatch-oriented occupations of public safety, transportation, and field services.

Other analysts are not so sanguine. They question whether

NEXTEL or any of the ESMRs can compete nationally with the major cellular carriers.

Frost & Sullivan/Market Intelligence believes ESMR will open up new markets for users not needing dispatch services, such as salespeople. This would, according to the international market research firm, position SMRs as a low-end cellular carrier, providing similar services at a lower monthly rate. F&S/MI foresees a larger base of users for SMRs, but not the annual doubling of their customer base over the next three years that some industry analysts are projecting. F&S/MI says there are too many potential variables, such as how well SMRs target users who currently don't subscribe to any mobile phone system, and how they approach those who already use cellular phones but express concern about the coverage provided by SMR. Those issues do not seem to have slowed the growth of SMRs. Based on a survey conducted by EMCI and the American Mobile Telecommunications Association, EMCI estimates SMRs had 190,000 subscribers at the end of 1993—a 14 percent increase over 1992. However, with the advent of digital service, EMCI projects ESMR subscriber growth will climb to 4.4 million in 1998. Although this is a significant projection, it is still well off the 20 million mark projected by the cellular industry by 1998.

Major cellular equipment suppliers, meanwhile, are experiencing the most exciting time in their history. A study by the U.S. International Trade Commission (ITC) shows that only four companies, Motorola, AT&T, Northern Telecom, and Ericsson, account for nearly 90 percent of global cellular network equipment sales.

At least part of the reason for their success is due to the way these companies develop and produce products. Interviews with cellular equipment manufacturers by the study's authors suggest that the breadth of product lines, radio manufacturing experience, integrated circuit (IC) designs, manufacturing competence, and advanced manufacturing techniques are the principal factors influencing competitiveness. Together, Motorola and AT&T account for 42 percent of global sales of cellular network equipment, although this is largely due to their predominance in the U.S. Outside the U.S., Ericsson appears to have a clear competitive advantage.

The ITC attributes Ericsson's success in foreign markets to

Honda offers this in-dash cellular telephone kit with a small antenna and a transceiver, including a keypad, display, and microphone, in a single unit.

three factors. The first is the company's ability and willingness to produce equipment for multiple system standards. Ericsson is the only manufacturer supplying substantial amounts of network equipment for the Advanced Mobile Phone Service (AMPS), the AMPS-derived Total Access Communication System (TACS), and Nordic Telephone (NMT) analog standards, while Motorola and AT&T have focused on the AMPS market. The apparent unwillingness of U.S. firms to produce equipment to the NMT standard has cost those companies market share in Asia, Africa, Eastern Europe, the former Soviet Union, and the Middle East, where the NMT system has won wide acceptance.

Second, Ericsson has developed and maintained core competencies in both switch and cell site equipment manufacturing. According to the ITC, Motorola "has seen its reputation as a manufacturer of high-quality switches deteriorate, reportedly due in most part to its inability to produce high-capacity switches." The third factor in Ericsson's favor is its global marketing of wireline telecommunications equipment, establishing long-time supplier relationships with overseas telecom service providers.

Ericsson's dominance in foreign analog cellular markets is likely to continue, according to ITC's analysis, but Motorola and Northern Telecom should be very competitive with Ericsson in overseas digital cellular markets. AT&T and Nippon Electric Corp. (NEC) of Japan match Ericsson in terms of their ability to supply complete cellular systems, but both may produce equipment only for U.S. and Japanese TDMA standards. The ITC also found that Motorola and Northern Telecom match Ericsson in terms of producing equipment for all digital standards, but industry representatives interviewed for the study "voice uncertainty" concerning the individual abilities of these companies to supply both cellular switches and cell site equipment for digital cellular networks. The ITC also rates Motorola and Northern Telecom as highly competitive when they go up against Ericsson in Western Hemisphere markets.

Motorola comes out on top in both cellular phone sales (it is the dominant global manufacturer of cellular phones with 23 percent of the world market) and cellular phone makers' semiconductor sales. With the exception of Nokia of Finland, Motorola's principal competitors are Japanese electronic firms. Matsushita, Mitsubishi, and NEC together account for 30 percent of the global cellular phone market. Motorola, Nokia, Matsushita, Mitsubishi, and NEC will likely remain world leaders in cellular phone manufacturing for the foreseeable future; however, the ITC expects their collective share of the global market to decline as new producers, especially those in Europe, increase production and sales.

Among all cellular phone manufacturers, only Motorola has double-digit market shares in the United States, Japan, and Europe. The ITC attributes Motorola's preeminence to its production of a broad range of cellular phone models.

More worrisome for U.S. cellular service providers is the lack of a single U.S. digital cellular standard, which may put U.S. cellular carriers at a competitive disadvantage in the global market. Some industry analysts believe that, because of its size, population, and rapid market growth, the U.S. cellular market can comfortably handle more than one digital standard—at least for the near future. Nevertheless, several U.S. companies have covered themselves by aggressively aligning themselves with foreign cellular partners to develop digital cellular systems in those areas.

More significant perhaps are the changes taking place in the market itself. Cellular phones are becoming smaller and cheaper, and they are beginning to fit more comfortably into traditional consumer electronics distribution patterns. It is not clear that Japan will dominate this market in North America as it has other consumer electronics markets. Europe's top telecommunications suppliers have been showing a new aggressiveness that may strengthen their position in this fast-paced market. The United States, however, led by Motorola and AT&T/McCaw, appears likely to prevail in virtually every area that counts—technology, product development, and marketing.

Based on its analysis of quarterly surveys of the cellular market, CTIA reports, and additional studies, Herschel Shosteck Associates, Inc., a specialist in cellular industry economics and market analysis, says that infrastructure investment and cell site construction should continue to grow in the U.S. through 1996. On the one hand, rural operators rushed to meet construction deadlines mandated by the FCC. In 1991, 501 new systems came on line. In 1992, only 254 new systems were activated. In 1993, it will be negligible. On the other hand, Shosteck says the number of cell sites continues to rise. In 1991, operators added 2,231 sites. In 1992, they added 2,460, reflecting continued expansion of established systems to fill dead spots in their systems and to add capacity.

Shosteck forecasts that total cells will more than double by year-end 1996. Capital investment in systems will show parallel growth. This means that future infrastructure investment will equal, and arguably exceed, the more than $2 billion per year spent in 1991 and 1992. While total investment continues to expand, cost per new cell is falling. In 1989, the cost of new cells peaked at $1,256,000; by 1992, it had fallen to $1,053,000, or by 16 percent. Shosteck, whose work is followed closely by the cellular industry, attributes this decline to smaller diameter cells inherent to maturing urban systems. These cells use smaller and less costly towers and, in some cases, no towers at all, and require less land. Based on these factors, Shosteck foresees a continued decline in cell site costs.

Meanwhile, cellular carriers are moving to microcells—systems formed by splitting existing cells into smaller geographic areas—to fill system dead spots and to provide service in hotels, airports, train stations, enclosed malls, and other heavy-traffic indoor areas. Microcells are also expected to be an important part of the infrastructure of PCS networks, which will use very low power transmitters and portable phones. Bell Atlantic Mobile, for one, is already installing a major microcellular system to preempt direct competition from PCS networks. But the driving force behind all this activity, Shostek believes, is not added capacity, but improved service coverage.

Price will also be critical for cellular carriers competing with PCS. Here, history is on the industry's side. Shosteck's data indicates that in December 1983 (just months after the first cellular system went on-line in Chicago), the average low price of a cellular telephone was $2,628. By year-end 1992, the average low price was $174. In 1983, 250 prime-time minutes per month cost $150. In 1992, equivalent use cost $127. Competition continued to heat up among cellular carriers, and prices continued to drop as the carriers began to look for new ways to promote cellular phone use. By mid-1993, Cellular One was offering a Motorola handheld "flip phone" for $99 and no activation fee.

Lower costs became the essential factor driving the growth of cellular subscribers. It will also be the essential ingredient in competing with PCS.

Chapter 3

Personal Communications Services

(Or Personal Communications Something)

Personal Communications Services, or PCS. What is it, exactly? No one seems to know, *exactly*. Whatever it is, the Federal Communications Commission (FCC) has authorized more than 170 companies to test it and to determine if the public will buy it. The FCC is also forcing very large users of microwave transmission systems, including public utilities, railroads, petroleum companies, and state and local governments, to move their communications operations to a new frequency band to make room for PCS.

By definition—the best one that anyone has come up with so far—PCS is not one service but a variety of emerging personal communications services. Former FCC Commissioner Ervin S. Duggan told an industry group in May 1993 that PCS "has yet to be clearly defined beyond a broad concept: a family or continuum of wireless communications services at lower power and with more, smaller cell sites than cellular telephones. Some services will be feature-rich and therefore are presumed to be more costly; other, more modest offerings will perhaps be less costly." Despite this vagueness, virtually everyone seems to agree that PCS has the potential to match or even surpass the extraordinary success of the cellular telephone industry, which Duggan admits "has grown beyond anyone's most expansive dream." (One broad-based demand study estimated that Telocator-designated personal telecommunications service telepoint and wireless business service would penetrate 30 percent of the U.S. population within 15 years of market launch.)

By most accounts, PCS will start showing up in campus-type environments, such as business and medical centers and industrial parks. This is what McCaw Cellular had in mind in 1992 when it created localized microcellular systems in New York to serve the Democratic National Convention and in Las Vegas for Comdex, the world's largest trade show. Using cellular handsets over low-power transmission systems, people at both events reported better battery life, resulting in longer talk time, and improved call clarity when compared with their cellular service.

Conceptually, the system introduced in March 1993 by Southwestern Bell Mobile Systems and Panasonic Communications & Systems Co. is very similar to McCaw's installations. Called the FreedomLink Personal Communications System and described by Southwestern Bell and Panasonic as the first commercially available PCS, it operates as an extension to any existing office phone system including a private branch exchange (PBX), Centrex, or key system. It uses cellular technology at very low power (about 10 mW) to communicate with the office telephone system. (Market research indicates that talk time is important to consumers and FreedomLink phones operate up to 5 hours in their PCS mode; talk time on the external cellular system is only $2\frac{1}{2}$ hours.) With a FreedomLink pocket phone, an employee can move throughout an office, factory, or business complex placing and receiving calls. Base stations, about the size of a door chime, are mounted on the wall throughout a building or complex and relay calls between the pocket phones and control unit. When the employee leaves the FreedomLink area, the pocket phone operates as a regular cellular hand-held phone using the external cellular system.

When a FreedomLink user receives a call, both the pocket phone and the user's desk phone ring. The user can either answer the call at his desk or from wherever the user is located, allow a secretary to answer it, or have the call directed automatically to voice mail. A visual indicator on the phone then alerts the user that a call has been received. FreedomLink isn't cheap; the cost varies among Southwestern Bell Mobile's markets, but ranges from $1,800 to $2,500 per extension depending on coverage area. By mid-1994, Southwestern Bell Mobile and its Cellular One divisions expect to market FreedomLink throughout their 54 cellular service areas.

Is this the type of product that's going to make the PCS personal communicator as ubiquitous as the personal computer by about 2010, as many analysts believe? If so, PCS is going to have to live up to its early billing as a low-cost service. If the results of studies by BellSouth hold up, that's going to be difficult. Market research by BellSouth indicates that wireless customers are willing to buy a service with fewer features than regular cellular, but only at a significant discount to cellular. At prices close to cellular, they want full cellular service. That's true, even though BellSouth gave the phone away for its market trials.

Where does that leave paging? Prudential Securities' telecommunications analysts, reporting in the Telocator *Bulletin*, the association's newsletter, say that based on their studies—in which they assumed that PCS would become commercially available in 1995, and that consumers would pay $55 per month for the service, while paging operators would charge $9.70 per month by 1995—PCS would not affect paging in the future any more than cellular has in the past.

At this early stage in its development, the market may be the least of the industry's problems. The larger struggle will be trying to unload all of PCS's regulatory baggage. The task is daunting.

In January 1992, the FCC proposed making available 220 MHz of spectrum for so-called emerging technologies within three frequency bands in the 1.85- to 2.2-GHz band (1850–1990 MHz, 2110–2150 MHz, and 2160–2200 MHz). Then, in July, the FCC said it planned to open up 20 MHz of spectrum (1910–1930 MHz) for a new class of low-power, user-provided wireless voice and data PCS called "User-PCS," which would cover a broad category of future applications, including wireless local-area communications for portable and desktop computers, wireless personal digital and messaging devices, and wireless office and home telephone systems.

One significant feature of User-PCS is that equipment would not have to be licensed by the FCC, but would only have to meet FCC-type approval requirements according to a "spectrum etiquette," a set of rules which would essentially allow the use of unlicensed digital radio devices as long as the same frequency band isn't being used at the moment by someone else within its range.

These proposals have not gone unchallenged. In this case, very large users of so-called fixed microwave services—the public utilities, railroads, and petroleum companies—may have to move from their current assigned band (2 GHz) to higher frequencies, to make room for the new wireless communications services. This could be a very costly process. But putting PCS anywhere else would result in unacceptable technical trade-offs in the power and size of PCS products.

The petroleum industry has estimated that 20,000 to 30,000 microwave stations are operating around the country in the 2-GHz band. The American Association of Railroads argued that private microwave systems will be a major part of the railroad industry's new $2 billion advanced train control system (ATCS). Law enforcement agencies questioned whether the emerging PCS industry was willing or able to compensate all the costs of moving current microwave users. The Utilities Telecommunications Council put the total cost to private industry and state and local governments to relocate existing microwave systems, assuming suitable alternatives were available, at somewhere between $4 billion and $7 billion.

Hoping to fend off the incursion of PCS providers, the utilities and other groups began an intense lobbying campaign in Congress. They won the immediate support of Senator Ernest Hollings (D., S.C.), the chairman of the Senate Commerce, Science and Transportation Committee, who introduced legislation that would force the commission to maintain the 2-GHz spectrum for existing microwave licensees. It wasn't until the FCC came up with a plan to prevent the disruption to incumbent microwave users that Hollings backed off. The new plan called for PCS to get the 2-GHz band, but allowed incumbent microwave users to move only on a voluntary basis.

Would-be PCS carriers continued to argue that they needed at least a portion of the 2-GHz band, and they needed it immediately if they were going to be competitive with the rest of the world. In June 1993, the FCC came up with what it hoped would be its final proposal for PCS spectrum use. The plan required PCS operators to pay incumbent microwave equipment users to move to new, higher frequencies. Under the FCC's new proposal, negotiations over the cost of the move would have to begin when the PCS provider filed an "application" requesting spectrum. For licensed users, the FCC allowed for two negotiating periods, starting with a fixed two-year period that begins with the commission's acceptance of applications for new services. If, during this period, negotiations are not successful, a one-year negotiating period would follow during which the PCS carrier could ask the FCC to order the eviction of the incumbents from their assigned frequency. For unlicensed devices, the FCC established a single one-year period beginning with the application of an unlicensed equipment supplier or incumbent licensee. In addition to financial compensation, the FCC authorized the use of tax certificates by incumbent microwave licensees to allow them to defer taxes on fees received from the newly ensconced PCS operators.

The only group of fixed microwave system users not affected by the FCC ruling are law enforcement and emergency medical service agencies; they will be allowed to continue using the 2-GHz band for public safety and medical purposes, but not for administrative use.

Historically, frequencies have been allocated through comparative hearings or by lottery. Telocator has generally opposed spectrum auctioning, fearing that the FCC may develop different rules and conditions for various classes of applicants, and that auctions may disrupt the ongoing licensing process. Several lawmakers also expressed concern about how to define the services covered in the auction legislation. Congress, however, favors spectrum auctioning. One reason is money: Government sources estimate the sale of frequencies to new communications services would generate more than $10 billion in revenue for the federal government. Some industry analysts place the value of new spectrum allocations to the government as high as $40 billion. Congress also believes that auctioning will speed the spectrum-allocation process, getting emerging technologies to market faster.

Whatever the value, the spectrum issue is absolutely critical to PCS providers. He who owns the spectrum owns the market. Consider this: In 1990, the Congressional Budget Office (CBO) estimated the market value of the spectrum of the entire broadcast industry (400-MHz worth), including television and radio, at $11.5 billion. At the same time, the CBO said the market value of the cellular industry's major metropolitan market areas alone (occupying just 50 MHz of spectrum) was $80 billion.

Most knowledgeable analysts believe that cellular carriers have a big jump on other would-be PCS operators—from four to five years by some estimates. In fact, some industry observers don't see PCS as even a long-term threat to cellular, suggesting that with the possible exception of highly localized networks, such as industrial parks and medical centers, PCS providers will have a tough time shifting current cellular user to less functional services.

That was one of the key messages from BellSouth's market trials. And it is key to the formation of a group of cable companies, equipment suppliers, publishers, and MCI that are trying to keep cellular carriers out of PCS, claiming that cellular's infrastructure offers an unfair advantage. Organized as the PCS Action Group, its members include American Personal Communications/Washington Post Co.; Omnipoint Corp.; Cox Enterprises, Inc.; Associated Communications, Inc.; Crown Media, Inc.; MCI Communications Corp.; Northern Telecom; Providence Journal Co.; Time-Warner Telecommunications; and Times-Mirror Cable Television, Inc. Its stated purpose is to promote the allocation of 40 MHz of spectrum for PCS. It also supports the assignment of only two licensees in each market area. But its real mission—indeed, the reason it was formed—was to gain sole access to the newly assigned PCS frequencies.

The CTIA responded, in part, by suggesting that if it is important to keep cellular out of PCS because its existing infrastructure gives it an unfair competitive advantage, then the cable companies and MCI should be kept out because of their existing infrastructure. The CTIA's pitch is that its members want to evolve in a competitive market, from their narrowband wireless service into new broadband wireless services, just as publishers and cable and broadcast companies are growing out of their core businesses. CTIA President Thomas E. Wheeler says, "Every one of these companies is seeking to keep cellular out of the new spectrum while expanding into that spectrum themselves, and essentially in the markets they already dominate with their other communications assets. They can't have it both ways—operating PCS spectrum in market can't be good for the goose and bad for the gander."

Several cable system operators already own and operate cellular networks and have received FCC approval to test PCS concepts, using cable system-derived schemes or by tying their cable and cellular systems together into a completely new service. A good example is Comcast Corp., a cellular and cable system operator, which has been testing a PCS in Trenton, New Jersey, using both technologies. Cable Television Laboratories, Inc. (CableLabs), a research and development consortium of cable TV system operators, is exploring whether there is a role for the cable television industry in this new technology. CableLabs is working with several companies, including Motorola, to integrate PCS and cellular radio systems with cable's broadband network. The CableLabs/Motorola project has focused on defining suitable cable network architectures for transmission of PCS baseband and RF signals using fiber optics and coaxial distribution technologies. Motorola is also working individually with several CableLabs members, including Comcast Corp. and Cablevision Systems, to confirm the potential value of using wireless communications (such as PCS) in cable's broadband network.

The FCC's task is not unlike the one it faced 10 years earlier when it established the ground rules for cellular. Four different PCS implementation plans are under consideration. One calls for the creation of 487 "Basic Trading Areas (BTAs)" or territories based on the Rand McNally Atlas. Plan B would involve the creation of 194 Local Access and Transport Areas (LATAs), the local telephone network areas controlled and operated by the regional Bell telephone companies. A third proposal calls for 47 "major trading areas" as defined by Rand McNally. Finally, the FCC is considering the establishment of one national market to be served by a single company or consortium. MCI, the nation's second largest long-distance company, and a member of the PCS Action Group, has already formed an alliance of more than 150 companies specifically to apply for a national PCS license—an alliance formed largely out of fear of the dominance in PCS and other wireless markets by rival AT&T and McCaw Cellular.

US WEST has submitted its own PCS plan to the FCC. It calls for one license per carrier per market, with no more than three PCS carriers per market. Each carrier would have at least 30 MHz of spectrum. The US WEST proposal, which is based on the company's experience with PCN in the United Kingdom, also includes recommendations for eligibility, spectrum, competition, and the size of service areas. For one thing, US WEST says the one license per carrier per market will encourage the efficient use of spectrum.

None of these plans makes sense to the cellular industry, which wants five PCS carriers per market with 20 MHz per carrier and a PCS structure that resembles the metropolitan service area/rural service area (MSA/RSA) scheme currently used by cellular carriers.

The CTIA admits that, on the surface, issuing licenses for 47 markets or 194 markets or even 487 markets would be better and faster than the 734 license areas needed to match cellular markets. But the appeal of the basic math doesn't hold up in terms of what has actually occurred in cellular licensing.

As the CTIA sees it, it took three full rounds of rule making by the FCC, extending over most of a decade, to establish and fine-tune the cellular service areas. The CTIA believes that using the cellular MSAs and RSAs would be faster because the cellular service areas are already established, are well known to the communications industry, and cover the entire country.

The cellular leadership also sees several problems with using the Rand McNally trading areas to license communications services. For example, as late as 1987, the FCC combined cellular RSAs 6 and 7 in Minnesota into a single market to ensure better service to the corridors of Highways 65 and 169 and to the Central Lakes area. According to the CTIA, those counties are separated among five Rand McNally Basic Trading Areas. And since the Rand McNally trading areas were not designed with telecommunications services in mind, they randomly intersect LATA boundaries and telephone numbering plan areas, unintentionally, according to the CTIA, creating the potential for years of waiver proceedings and delay.

As for LATAs, the CTIA says these include only the local landline exchanges of the Bell operating companies and not the rest of the telephone industry. Moreover, the LATA boundaries are constantly changing—99 times since 1983. Also, the FCC has no jurisdiction over LATA boundaries, making a PCS licensing scheme based on LATAs a potential regulatory landmine. As for a national license, the cellular industry association says it "does violence" to the FCC's goals of speedy deployment, diversity of services, and competition.

Clearly, the stakes are high and the FCC is under some pressure to take action on PCS. The situation is not unlike the very tight schedule Congress gave the FCC to solve some of the cable television industry's problems, but the commission came through with flying colors. Even the cable industry was impressed. As a result, Congress has set a timetable for FCC action on PCS.

After more than two years of intense debate over whom should have access to spectrum for "emerging technologies," how much spectrum should be made available for these new services, and how to award that spectrum, the FCC has formally authorized PCS in the 2-GHz band. Under the FCC's new rules, as many as seven new wireless services may become available in every major U.S. market. The auctions are scheduled to begin in May or June 1994.

Specifically, the FCC allocated a total of 160 MHz at 1850–1970 MHz, 2130–2150 MHz, and 2180–2200 MHz for PCS—four times the spectrum originally allocated for the cellular telephone service in 1993. The decision calls for 120 MHz to be allocated for licensed PCS (1850–1890 MHz, 1930–1970 MHz, 2130–2150 MHz, and 2180–2200 MHz), and 40 MHz for unlicensed PCS devices (1890–1930 MHz).

Cellular carriers may bid on any PCS license outside of its cellular service area "or in any area where the cellular licensee serves less than 10 percent of the population of the PCS service areas." But a cellular carrier will be allowed to bid on only one of the 10-MHz BTA licenses in its cellular area. The 10 percent rule is designed to keep cellular carriers out of the largest markets because they already serve more than 10 percent of the population in those markets.

Specifically, the licensed allocation was channelized into two 30-MHz channel blocks, one 20-MHz channel block and four 10-MHz channel blocks. The PCS areas adopted were Major Trading Areas (MTAs) and Basic Trading Areas (BTAs), generally defined by the Rand McNally Atlas. There are 51 MTA and 492 BTA-defined service areas under the plan adopted by the FCC.

The allocation by channel blocks, frequencies, and service areas break down as follows:

Channel Block A: (30-MHz), 1850–1865/1930–1945 MHz, MTA;

Channel Block B: (30-MHz), 1865–1880/1945–1960 MHz, MTA;

Channel Block C: (20-MHz), 1880–1890/1960–1970 MHz, BTA;

Channel Block D: (10-MHz), 2130–2135/2180–2185 MHz, BTA;

Channel Block E: (10-MHz), 2135–2140/2185–2190 MHz, BTA;

Channel Block F: (10-MHz), 2140–2145/2190–2195 MHz, BTA;

Channel Block G: (10-MHz), 2145–2150/2195–2299 MHz, BTA.

The unlicensed allocation was channelized into two 20-MHz blocks, one for voice services and one for data. The FCC said it took this approach to ensure a robust and competitive market for PCS and permit broad participation in the provision of PCS, including participation by small businesses, rural telephone companies, and businesses owned by minorities and women. The FCC also set the licensing term at 10 years, which is similar to current cellular service requirements.

Where does this leave MCI, which has organized a consortium of close to 200 companies to create a nationwide PCS service? The FCC did not specifically allow for a single nationwide license; however, it did suggest that a company could submit a sealed bid that combines every regional license into a single license request. That could be very expensive: To be considered for a national license, the bid would have to exceed all the individual bids from each region. The bidder would also have to agree to make service available to one-third of all potential customers within five years and to two-thirds within seven years. MCI said it would have to study its options.

To deal with the issue of interference between unlicensed PCS devices, the FCC adopted the "spectrum etiquette" guidelines proposed by the Wireless Information Network Forum (WINForum). The basic idea behind the WINForum plan is to "listen before you talk." In other words, unlicensed PCS devices—both voice or data—would not transmit if the spectrum it occupies is already in use within its range.

Still, many questions remain. Will PCS be so much better than cellular? Clearly, if PCS is going to succeed on its own, consumers will demand much higher performance from PCS than cellular and other wireless personal communications services. That means a highly reliable and flexible system with smaller phones, improved battery life and range, and lower cost.

Cellular carriers are already offering PCS-like services in some areas allowing subscribers to use a single personal phone number to reach anyone virtually anywhere. Cellular phones themselves are getting smaller all the time, and the emergence of seamless national networks will give them just about all the range most subscribers will need.

Will PCS actually be cheaper than cellular? Few analysts are convinced that it will be cheaper. Cellular phone and airtime prices are dropping; a U.S. General Accounting Office study indicates that rates for cellular airtime declined 27 percent from 1985 to 1991, and that the average monthly bill dropped from $96.83 in 1987 to $68.68 in 1992. Some carriers are even giving phones away. McCaw Cellular has already reduced its prices by 20 percent for users of its new digital service between Florida and the Pacific Northwest. NYNEX has cut its cellular phone prices in half and offers two months of basic service for $29.95 a month with no activation fee—a savings of $50.

Competing with cellular is only a part of the problem facing would-be PCS providers; first, they have to build their networks and that's going to be very expensive. Starting with the cost of acquiring a license through auction, which could be as much as $2 billion, PCS operators must then string thousands of transmitters every 500 to 1,000 feet to create a viable network. A study by Arthur D. Little indicates that a network serving 7 million subscribers in a 300-square-mile area would cost about $6.3 billion.

The cellular carriers aren't waiting around. They're already experimenting with several new and innovative services using different technologies at different price points on their way to developing new PCS networks.

McCaw Cellular and Tele-Communications, the largest cable television company in the U.S., have combined their cellular and cable technologies into a residential microcellular service in Ashland, Oregon. Promoted as the Ashland Personal Telephone Service, McCaw and TCI gave 200 area residents pocket-size, battery-powered telephones to use virtually anywhere in their community, including homes and office buildings. Using a modified version of the conventional cellular "flip phone" to communicate with four low-power microcells operating on standard cellular frequencies, subscribers can also receive conventional cellular service outside Ashland. The microcells are interconnected by TCI fiber-optic cables and McCaw's cellular switching facilities.

Bell Atlantic Mobile Systems has taken a somewhat different approach in a PCS trial in the Baltimore/Washington, D.C., area. This new service allows Bell Atlantic Mobile customers to use a single phone number to receive calls at their office or home on a cellular phone, a facsimile machine or pager, or by a message left on voice mail. In operation, calls to that one number are received by the Bell Atlantic Personal AccessLine computer, which handles them in accordance with the customer's instructions or a prearranged schedule. For example, incoming calls may be directed to an office during the day, to a cellular phone after 5 P.M., and to a home after 6 P.M.

Bell Atlantic and Motorola are jointly testing a cellular-based PCS using Motorola's Personal Phone Service 800 (PPS 800) system and new pocket-size Personal Communicator telephone. The trial, in Pittsburgh, allows users with a single personal telephone number to be reached at home, in the office, and on the road. The PPS 800 system uses Motorola's Narrowband Advanced Mobile Phone Service (NAMPS) to minimize the frequency-allocation impact on the existing macrosystem. It is designed to take advantage of the wired and wireless networks already in place, including Bell Atlantic's Advanced Intelligent Network provided by Bell of Pennsylvania in Pittsburgh. The Motorola handset operates as an extension of a PBX or Centrex system in the office and as a cellular phone when the user is mobile, whether walking or riding in a vehicle.

QUALCOMM's trial is aimed at demonstrating the feasibility of using its frequency-efficient CDMA technology in the 1850–1990-MHz band for PCS.

Cellular carriers feel very strongly about their approach to PCS. They believe that their customers want to be able to define their own unique set of features and functions, and that they want to be able to make and receive calls any time and any place. They want smaller portable handsets, and they will demand ubiquitous coverage.

These assumptions, made in late 1991 and early 1992 when most PCS trials were just getting underway, have pretty much held up in the results of the first PCS market trials report by Telocator, the Washington, D.C.–based personal communications industry association.

According to the Telocator report, compiled from 11 trials and one market study, broad coverage is the most important quality sought by consumers. In fact, reaction to coverage was so strong that it was found to have a direct effect on demand. One study concluded that broader coverage could increase demand by as much as 40 percent, while reduced coverage could depress demand by as much as 32 percent. The coverage finding was part of a larger pattern reported by numerous trial surveys showing that, in addition to large coverage areas, users want two-way calling and the ability to use a service while driving, all at a price lower than cellular service. Trial findings also suggest that consumer education on attributes of PCS such as coverage, features, and battery life, in conjunction with market introduction of new services, will play a vital role in early mass market acceptance of, and consumer satisfaction with, new PCS offerings.

Quality was another concern. Coverage proved less reliable in hilly areas, and concrete and steel buildings were more difficult to penetrate than wooden structures. Nearby buses and trucks also caused significant radio frequency (RF) interference.

Comparing this with Telocator's earlier estimate of 60 million new users of PCS of all types by 2002, the association says, "...the reports are consistent in showing that there is a tremendous demand for personal communications services." The report also concludes that the introduction of PCS will increase the penetration of all wireless services—just as cellular did for paging.

Among the user benefits identified by trial participant were safety and security, convenience, and family coordination. Another trial reported the key benefits of wireless/PBX/wireless Centrex to be minimization of telephone tag and quicker resolution of communications.

Some of these trials began as early as 1990, although several of them were started in 1993. The trials vary in size from 27 to more than 3,000 participants; they involve residential and business customers. Some trials do not charge users. However, most are testing a broad range of pricing options.

Companies participating in the trials included American Personal Communications, Ameritech, Bell Atlantic Mobile, BellSouth Enterprises, GTE, Rogers Cantel Mobile, and Telesis Technologies Laboratory.

In the American Personal Communications' telepoint trial in Washington, D.C., and Baltimore, for example, home base station and handset rentals along with the monthly service charge were $15 per month plus 13 cents per minute of usage in the public service areas. With the home base station, the handset offered two-way communication at home and one-way outgoing calling in the public service areas. Service areas include Capital Hill and the business district in Washington, D.C., the Inner Harbor business and residential district of Baltimore, as well as Dulles and National Airports. Midway through the trial, participants were provided with a handset that integrated a pager, for an additional charge of $7 a month.

In BellSouth Enterprises' Orlando PCS trial, DriveAround PCS (a two-way fully mobile service) was priced 15 percent below cellular and offered a ubiquitous service coverage area throughout central Florida, including Orlando, Melbourne, and Daytona. WalkAround PCS (a two-way service with no handoff) was priced 30 percent below cellular, with ubiquitous service coverage area throughout central Florida. OutBound PCS (a one-way service with no handoff) with a pager was priced 50 percent below cellular and offered a ubiquitous service coverage area throughout the metropolitan Orlando area.

In Cantel Personal Communications' Ottawa PCS trial of telepoint and advanced telepoint (figures are in Canadian dollars), the basic package was \$7.95 per month, plus 60 cents per 5-minute call and 20 cents per minute thereafter. The per-call package included 15 calls for \$14.95, plus 40 cents per additional 5-minute call and 20 cents per minute thereafter. And the unlimited package included 350 minutes for \$49.95, with additional usage cost of 15 cents per minute.

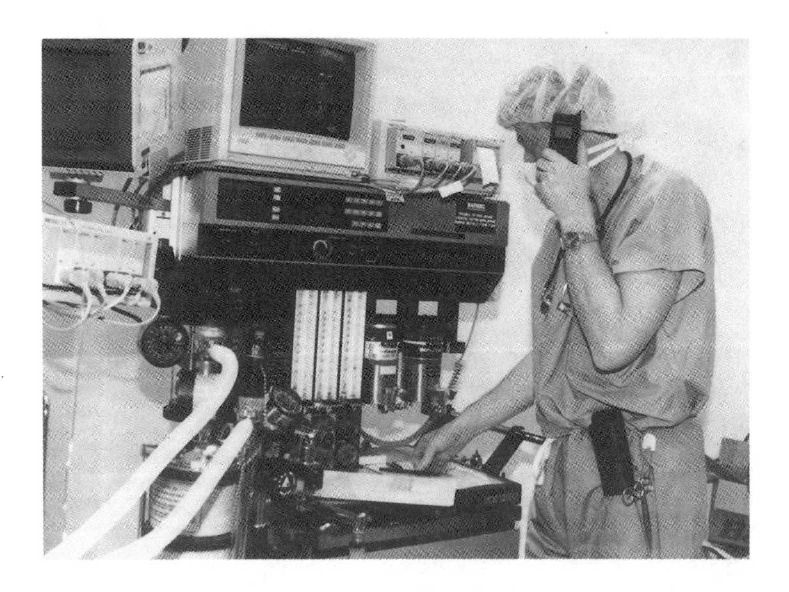

The FreedomLink PCS-concept phone has been field tested at New England Deaconess Hospital. It allows the hospital's medical staff to communicate without pagers and overhead speakers, even in the operating room.

In GTE's ongoing Tele-Go PCS market trial in Tampa and St. Petersburg, Florida, the replacement of home telephone service with PCS cost $40 per month for 800 minutes in the "home area" or home neighborhood. Calls in the "premium area," which includes the greater metropolitan area of Tampa and St. Petersburg, cost 15 cents a minute. Add-On to Existing Telephone Service costs $20 per month and includes 400 minutes in the home area; calls in the premium area cost 25 cents per minute. In GTE's telepoint service, replacement of home telephone service includes 800 minutes in the home area for $32 for advanced telepoint service; telepoint service is offered in premium areas and calls are 12 cents a minute. Add-On to Existing Telephone Service costs $16 per month for 400 minutes in the home area for advanced telepoint; telepoint service is offered in the premium area with calls priced at 20 cents per minute.

Telesis Technologies Laboratory is testing over 10 service concepts in its San Diego trial with 1,500 trial participants using different coverage and pricing options.

Two trials concluded that to attract users, total monthly charges for telepoint/advanced telepoint should be significantly less than cellular, averaging between $25 and $50, ideally in the lower end of that range. One study found customers to be more comfortable with per-call pricing, while another found a strong preference for a flat monthly charge similar to some land-line service.

In one telepoint/advanced telepoint trial, the average call length was 2 minutes, similar to the average cellular call. In limited coverage areas, usage increases as users become more familiar with base station locations and coverage areas. Campus environments, where the user is certain to be in range, have the highest usage.

The trials have also revealed that services offering limited coverage areas, such as telepoint, may benefit by adding a pager to the service offering. Several trials integrated pagers into a one-way telepoint service offering to allow two-way communication. One trial found current pager users more likely to subscribe to a one-way telepoint service with an integrated pager. Requested handset features include a calling number display and an in-range indicator; requested base station features include a keypad and a speakerphone.

Each of the reporting companies in the Telocator study plan to continue their trials with more emphasis on marketing and technical issues, particularly pricing, ergonomics, concept testing, and marketing communications. Technically, the focus of ongoing trials will be on code-division multiple-access (CDMA), coverage, and intelligent network features.

Internationally, the Japanese are conducting PCS network trials based on the digital cordless standard known as the Personal Handy Phone (PHP) operating at 1.9 GHz. The Europeans are moving toward personal communications networks (PCNs) via microcellular networks based on the digital cellular standards such as the Global System for Mobile Communications (GSM), operating at 900 MHz, and digital communications service (DCS) 1800, operating at 1.8 GHz. (See Chapter 6, "International Cellular/PCS Markets.")

Meanwhile, six would-be PCS providers—Bell Atlantic, BellSouth, Pacific Bell, Stentor (the administrative arm of Canadian wireline cellular companies), Time Warner, and US WEST—have joined together to push for technical standards and open interfaces leading to an industrywide PCS system architecture. In the view of the group, which calls itself the PCS Technology Advocacy Group (PTAG), the absence of technical PCS standards is impeding development of the industry. The PTAG hopes to work with manufacturers and established standards-setting bodies to help speed the development process. In addition, US WEST will provide the Boulder PCS testbed facilities as a common testing facility.

Chapter 4

Mobile Satellite Services

Huge (Potential) Market, Huge Risk

Arthur C. Clarke's *2001 Space Odyssey* stories have been tremendously popular throughout the world, and a huge financial success. But they don't have quite the same impact as another Clarke idea, one offered more than a quarter century ago: Create a system of uninterrupted communications through a global network of satellites, circling the earth in geostationary orbit. Since then, more than 320 satellites have been launched into orbit, and satellite communications has become a $10 billion industry.

Mobile satellite services (MSS)—an industry that barely existed in 1989—represents one of the next decade's fastest-growing telecommunications markets. The number of communications satellites circling the earth could almost double, and the industry's revenue could grow 10-fold within a decade if all the mobile satellite systems currently being proposed are launched on schedule. Ultimately, The Gartner Group said in a study commissioned by NASA, mobile satellite services "will be in the same league as personal computers."

Market projections for MSS equipment manufacturers and service organizations range from $1 billion by 1995 to $9 billion by the year 2000. The International Maritime Satellite Organization, or Inmarsat, the London-based 67-country member satellite consortium that provides data links for ships, aircraft, and other mobile users, expects the number of land-based switching stations, or gateways, compatible with its own system to climb from 50,000 to more than 400,000 by the turn of the century.

Initially targeted at developing countries with little or no telecommunications services, the MSS market is already finding its way into more commercial and consumer applications, such as airlines (there are nearly 5,000 commercial aircraft, most of which will eventually have in-flight, satellite-based public telephones), pleasure boats (at least 10 million in the United States alone), and recreational vehicles (8 million at last count). Hand-held dual-use (cellular/satellite) phones will also become increasingly popular with hikers, campers, and others engaged in recreational activities. But it is the mass business/consumer market that is creating most of the excitement.

Virtually from the launch of the first communications satellite, satellites have been maintained in geostationary orbit, 22,300 miles above the earth's equator. The satellite moves at a velocity that makes it appear to stand still over a fixed point above the equator. In operation, each satellite produces a series of beams that divides its coverage into a pattern of overlapping cells, or "footprints," on the earth's surface. The total area visible to the satellite usually includes one or more regions that are usually heavily populated. As a result, satellite antennas are designed to provide coverage to only a portion of the total area visible to the satellite. In some systems, the antennas can be steered to provide service to specific areas over extended periods of time.

But this type of orbit causes problems. Because of the satellite's distance from earth, it takes at least a half-second for the signal to bounce between ground stations and satellites. The time delay can be frustrating for voice communications; you might be responding to a question while the person at the other end is already asking the next question. Also, the distances at which geosynchronous satellites must operate requires technical capabilities that make them very costly, ranging into the hundreds of millions of dollars.

Geostationary satellites also have certain advantages. For one thing, satellite communication carriers don't have to actually own a geostationary satellite. They normally lease channel capacity on a satellite and sell communications services, usually to niche markets, such as the long-haul trucking industry. If they need more capacity, they simply buy it. As a result, they don't incur the high cost and risk required of launching and maintaining their own satellite system.

American Mobile Satellite Corp. (AMSC) expects to launch the first geostationary mobile communications satellite. AMSC petitioned the FCC for a license to build and operate a mobile satellite system in February 1988. Just over a year later, the FCC approved AMSC's license for the construction, launch, and operation of the first U.S. mobile satellite system. Hughes Aircraft and Spar Aerospace Ltd. of Canada will build the first of AMSC's three satellites, and General Dynamics Commercial Launch Services and Westinghouse Electric Systems Group will integrate the ground network and supply voice and data terminals for the spacecraft. Mitsubishi Electric and Westinghouse Electric will produce AMSC's cellular/satellite telephones.

Rather than wait for the launch of its own satellites, AMSC is leasing capacity from Inmarsat and is providing mobile data and position-location services to the transportation, maritime, rail, and remote-monitoring markets. By late 1994, AMSC hopes to be offering mobile voice, data, fax, paging, and position-location services throughout the United States, including Alaska, Hawaii, Puerto Rico, and the Virgin Islands.

The company has already signed distribution agreements with 60 cellular carriers throughout the U.S. Collectively, they serve 633 markets representing a population of 152 million. As AMSC distributors, the cellular carriers will offer their customers, for the first time, access to mobile communications service anywhere in North America through a cellular/satellite mobile phone. When the customer is in an area covered by cellular service, the phone will route the call through the available cellular network. Only when no cellular system is available will the phone automatically process the call through AMSC's earth stations and public-switched telephone network. AMSC's goal is to sign cellular carriers in each of the FCC-designated metropolitan and rural service areas by the end of 1993. In time, AMSC hopes to sign most cellular carriers in the U.S. to its cellular/satellite service.

One of the first to sign on was McCaw Cellular Communications, which owns a 32 percent stake in AMSC. Other shareholders include Hughes Communications Inc. and Mobile Telecommunication Technologies (Mtel), each with a 29 percent share in

AMSC, and Singapore Telecom and several vendors, including General Dynamics and Westinghouse Electric, holding the remaining 10 percent.

Telesat Mobile Inc. (TMI) of Canada hopes to build a satellite system that is virtually identical to AMSC's to provide mutual backup and seamless MSS for all of North America. However, Telesat Mobile has been seeking protection from creditors pending a possible reorganization. In March 1993, the company reported debts of about $111 million and reported that it hasn't been able to raise additional funding to meet the rising costs of building a mobile satellite communications network. Telesat Mobile is owned 80 percent by Telesat Canada, which operates Canada's domestic satellite network, and 20 percent by Itochu Corp., a Japanese trading company. Telesat Mobile's biggest creditor is Telesat Canada. As a result, Telesat Canada isn't expected to provide Telesat Mobile with any additional funding. Telesat Mobile's second largest creditor is Arianspace, the French space organization, which has a contract to launch Telesat Mobile's satellite in 1994. Telesat Mobile has asked its creditors to approve a plan that calls for BCE Inc., a Montreal-based holding company, to take over the satellite organization. AMSC is totally independent of its Canadian counterpart and its financial problems, but still hopes to use Telesat Mobile as a backup system.

Several new mobile satellite services are taking an entirely different approach. Rather than buy capacity on geostationary satellites, they're developing and plan to launch a constellation of new, smaller, and much less expensive satellites that will operate in low earth orbit (LEO). The altitude is critical. Too high an orbit would require a very large rocket and result in huge launch costs. It would also put the satellites in a radiation environment that would drive up satellite-component costs. Too low an orbit would add to fuel requirements needed to maintain orbit, increasing operational costs.

In operation, LEO satellites will communicate with small, battery-powered, cellularlike portable telephones. Ground stations will link the satellites to the public-switched telephone network.

At least 13 LEO mobile satellite services have been proposed—five so-called "Big LEOs" and eight "Little LEOs." The differences are significant. Little LEOs will operate below 1 GHz and will offer a limited range of communications services, primarily position-location and two-way communication of very brief text messages. Big LEOs operate above 1 GHz and will offer their subscribers voice, paging, facsimile, data, and radio determination satellite services (RDSS) for navigation and tracking applications. A Big LEO phone could initially cost $2,500.

At that price, what's the advantage over the terrestrial-based cellular telephone system? It's a matter of positioning. Motorola is promoting Iridium, the most technically and financially ambitious of the proposed Big LEOs, as a premium satellite network for traveling businesspeople, as well as a backup or emergency service for cellular. Iridium could also provide basic telephone service in areas where there is no telephone service. Russia, for example, has a population of 250 million, but only 10 million telephones. In India, there are tens of thousands of villages without telephone service. Third World countries without telephone infrastructure could use Iridium with subsidized, solar-powered, centrally located telephone "booths" in every town and village. In areas of the world where mobile service is only provided through geostationary satellites, Motorola says Iridium could provide more channels, shorter delays, and worldwide networking.

Iridium was originally conceived as a 77-satellite system; hence, the name Iridium for the element whose atom has 77 orbiting electrons. Motorola engineers have since figured out a way to cover the same land mass with 66 satellites, configured in six polar-orbit planes of 11 satellites each. (Motorola says it has no plans to change the name to Dysprosium for atomic number 66.) The satellites, which will orbit about 420 nautical miles above the earth's surface, are to be phased, so that odd-numbered planes have satellites in corresponding locations, with satellites in the even-numbered planes staggered approximately midway between. The satellites will essentially travel in co-rotating planes up one side of the earth, cross over at the pole, and come down the other side of the earth. The earth, of course, will continue to rotate beneath them. Each Iridium satellite will transmit 48 beams; each beam can handle about 230 calls.

Motorola has developed a gateway to interconnect the Iridium constellation with the public-switched telephone network. The gateway is designed to handle such functions as call setup, caller location, and billing. The portable subscriber unit, when turned on, will link with the nearest Iridium satellite, which will then

Motorola's Iridium™ Dual-Mode Portable Telephone is designed to work with existing cellular systems by searching first for a local terrestrial channel before opening one of its own satellite-based circuits.

update the system's location register. Obviously, caller location is critical: An Iridium subscriber who lives in Seattle could carry his phone to Hong Kong or Tel Aviv and still make and receive calls through the Iridium network.

When Iridium was introduced in June 1990, Motorola estimated that it would cost $2.1 billion and would be in full operation in 1996. Since then, the cost has climbed to $3.7 billion and the program has yet to be fully funded. In October 1992, Motorola decided to sell 85 percent of the common stock in Iridium, Inc., the Washington, D.C.–based unit set up to manage the mobile satellite network. The stock sale was to occur in private transactions with both U.S. and foreign telecommunications companies. Motorola hoped to complete the stock transaction by April 1993 by selling shares to no more than 20 companies. The companies would share in Iridium's profits and might be allowed to buy ground stations that will link the satellite network with handheld cellular phones.

Motorola even set a deadline for itself: Fund Iridium by December 15, 1992, or seriously consider dropping the program. As the deadline passed, Motorola would say publicly only that it had received letters of intent from several potential investors and that, based on this level of interest, the company would proceed with the next phase of the initial round of financing. Initially, Motorola declined to identify specific investors, citing nondisclosure agreements. But word spread throughout the satellite community that the Brazilian government and United Communications Co. of Bangkok, Thailand, both planned to buy a 5 percent stake in the project, each valued at about $80 million.

In late January 1993, representatives from 18 potential investors met for the first time as a group in Geneva, Switzerland, to review the status of technical, marketing, and financial plans for Iridium. They signed subscription agreements or letters of intent, which represented an investment level beyond the $800 million targeted for the initial round of financing. But the agreements were nonbinding, forcing Motorola to set a new schedule for launching the Iridium system.

Then, in August 1993, Motorola said it had completed an initial $800 million equity placement for the program and identified the initial members of its Iridium consortium. Motorola will retain about a 34 percent equity interest in Iridium at a cost of

$270 million. Other investors include Nippon Iridium Corp., an alliance of 18 Japanese companies, and two Saudi Arabian groups led by the Mawarid Group. Companies with smaller shares in the program are BCE Mobile Communications/BCE Telecom International Inc. of Canada; Sprint Corp. of the United States; STET-Societa Finanziaria Telefonica per Azioni, an Italian telecommunications concern; United Communications of Thailand; and The Great Wall Industry Corp. of China. As part of its arrangement, BCE Mobile will operate and manage Iridium services in Canada.

The Japanese consortium is led by DDI Corp., Japan's largest private long-distance telephone carrier. Nippon Iridium Corp. plans to invest some $130 million immediately and another $130 million in 1994. DDI and its regional cellular phone subsidiaries, which use Motorola cellular equipment almost exclusively, will hold 49.8 percent of the new venture's shares. Other investors in Nippon Iridium include Kyocera Corp., a world leader in electronic materials and semiconductor packaging (10 percent); Sony Corp., SECOM Co. Ltd., and Ushio Inc., each with 5 percent; Mitsui & Co. Ltd. with 5 percent, and Mitsubishi Electric Corp. with 2 percent. Four financial institutions, the Sanwa Bank Ltd., Daiwa Securities Co., Industrial Bank of Japan Ltd., and Long-Term Credit Bank of Japan Ltd., hold the remaining shares.

Additional funding has come from Pacific Electric Wire & Cable Co. Ltd. of Taiwan, which purchased Iridium shares from Motorola and assumed the remainder of a $40 million equity commitment by Motorola in Iridium. The commitment represents a 5 percent share in Iridium.

Motorola also found an angel in, of all places, Russia. Although current plans call for McDonnell Douglas Corp. to launch most of the Iridium satellites, Motorola has signed a contract with the Russian Federation's Khrunichev Enterprise to launch a portion of the Iridium system constellation of LEOs from the Plesetsk cosmodrome in northern Russia. In return, Russia will spend $40 million on the satellite project, which would be recouped from future hard currency income.

Lockheed and Raytheon Co., which will produce antennas for the spacecraft, have a combined interest of less than 5 percent. Lockheed Missiles & Space Co. has signed a contract valued at more than $700 million with Motorola's Satellite Communications Division to provide the spacecraft bus for 125 Iridium satellites, systems engineering support, satellite vehicle assembly, integration, and test support.

Iridium, Inc., will continue to seek several additional investors and plans another $800 million equity placement in about two years.

Motorola now plans to launch its Iridium satellites beginning in 1996 with initial commercial service anticipated in late 1998. The business plan anticipates 75,000 subscribers by the year 2000; about 10 percent of those are expected to be Japanese. Nippon Iridium has already established a basic service charge of about $50 a month, plus $3 per minute. But at least one glitch remains for Motorola's Japanese partners: Under current Japanese telecommunications laws, domestic and international carriers operate separately and Nippon Iridium, or DDI, will have to convince the Ministry of Posts and Telecommunications (MPT) to change the law to accommodate the new service. Meanwhile, Japan is considering developing its own nationwide mobile satellite system. The MPT has formed a group to study the technical aspects of such a system, which probably would use the new N-Star satellite scheduled for launch by NTT in 1995.

Iridium will be expensive to operate. The satellites themselves will cost roughly $30 million each, and they're expected to maintain their orbit for only about five years. As a result, 16 of the satellites will have to be replaced annually. This represents a $480 million annual investment by Motorola and its partners, and those costs could climb with further delays or the onset of technical problems. Motorola has already spent more than $100 million to develop and test its new satellite system. The company estimates that it will need about 500,000 subscribers worldwide to break even financially, while 870,000 users would "represent a good business."

Some industry analysts remain skeptical. Iridium has no certain market, and it's not yet clear that Motorola can obtain service rights worldwide. How will international telecommunications authorities, who fear being cut out of satellite service revenues, respond to competition from a well-financed international consortium? Will Iridium, or any other MSS, have to pay for the use of ground stations based outside the United States? Who will control the ground stations? How will the ground stations in Mexico or Canada be treated under the North American Free Trade Agreement (NAFTA)? If Motorola wins the spectrum it needs for Iridium through an auction process, will international authorities ask for some form of tribute for similar consideration? When will Motorola receive FCC approval to launch Iridium?

Other Big LEO entrants don't appear to have been as aggressive in pursuing the MSS market as Motorola, so they haven't received as much attention. However, like Motorola, they have spent a lot of time and energy raising money to get their systems off the ground.

The TRW Space & Electronic Group's 12-satellite network, called Odyssey, is actually a medium-earth-orbit (MEO) system, designed to orbit the earth at an altitude of about 1,800 miles. TRW believes the MEO approach is more economical than LEOs because it would allow the Odyssey satellites to stay aloft longer, perhaps up to 7.5 years, compared to only 5 years for Iridium satellites. Odyssey will cover each region of the world with only two satellites, with each providing 2,300 telephone channels for voice, data, radiolocation, and messaging services. Like Motorola, Odyssey will provide a communications link between mobile subscribers and the public-switched telephone network.

QUALCOMM's satellite-based OmniTRACS system allows long-haul truck dispatchers to maintain two-way contact with their drivers, as well as track the positions of their trucks.

Odyssey is targeted at three market segments: institutional users, including government agencies; business travelers; and residents of sparsely populated regions who may never receive cellular service because there are not enough subscribers to pay for the infrastructure. If Odyssey wins FCC approval and actually flies, TRW plans to wholesale its services to other operators, rather than providing services to end users.

Loral Qualcomm Satellite Services, Inc., a joint venture of Loral Corp. and QUALCOMM, Inc., have proposed Globalstar, a network of 48 600-pound satellites in eight 750-nautical-mile orbital planes. Loral Qualcomm hoped to be in operation by 1997, but it is running behind schedule. Globalstar's management anticipated FCC authorization by the end of 1992 with construction beginning in January 1993. It missed those milestones, but it still expects its consortium partners to come through with the funding it needs to fully develop the system. Loral, and Space Systems/Alliance (SSA), the international satellite partnership backed by Aerospaciale, Alcatel, and Alenia, are major participants in the design, development, and production of the Globalstar satellites, as well as the ground station and operations segment. SSA is not an investor in Globalstar, but it would like to be. That's a problem with the FCC, which has considered adopting rules that would hold foreign ownership of U.S.–based telecommunications companies to 20 percent.

Two other Big LEOs are Constellation Communications, Inc.'s Aries and Ellipsat Corp.'s 12-satellite Ellipso system. In August 1993, Ellipsat and Constellation Communications received authorization from the FCC to launch four experimental satellites to validate their low-earth orbits. The Ellipso system was designed to serve the Northern Hemisphere, but it will also offer complementary services in the Southern Hemisphere through an additional constellation. Fairchild Space and Defense Corp. has agreed to be Ellipsat's prime contractor, responsible for in-orbit delivery of its satellites.

The one system that could present the most problems competitively for everyone is Inmarsat's Project 21, or Inmarsat-P (as in portable) program. Inmarsat has several options: It could use its vast, in-place geostationary satellite network to offer a new service very similar to Iridium. Or it could develop and launch a new Big LEO network. Inmarsat has already said that Project 21 would be a relatively low-cost system—low enough to be competitive with cellular—and would offer paging, facsimile, data, and RDSS in addition to voice service. Inmarsat could also use its new digital satellite communications Inmarsat-M service as a testbed for Project 21. Inmarsat-M, which is marketed in the U.S. by the Communications Satellite Corp. (Comsat), offers two-way voice, data, and fax services through lightweight, briefcase-size communication terminals. Comsat, with a 25 percent stake in Inmarsat, is the organization's U.S. signatory and largest shareholder.

Inmarsat has spent close to $10 million for six separate studies to determine the size of the global MSS market and what type of satellites and networks are best suited to the MSS task. Inmarsat assigned two teams to investigate its geostationary option—GE Astro Space/Matra Marconi Space and Hughes Space & Communications/British Aerospace/NEC Corp. Intermediate-orbit studies were turned over to TRW and the GE Astro Space/Matra Marconi Space team. Inmarsat assigned a European alliance, including Aerospatiale and Alcatel of France, Alenia Spazio of Italy, and Deutsche Aerospace of Germany, to study the LEO option. Inmarsat also hired Touche Ross, an international management consulting company, and Schema, a specialist in cellular market research, to study the MSS market. Schema and Touche Ross have prepared detailed country-by-country demographic and communications usage characteristic studies for eventual use by Inmarsat signatories.

Inmarsat has also been working with Nokia Mobile Phones, Europe's leading producer of cellular phones, to "explore the potential of satellite mobile telephones and related technologies." Nokia hopes to develop, manufacture, and market the dual-mode (cellular/satellite) Inmarsat-P phones. With an Inmarsat-P phone, calls could be made to or received from any phone anywhere in the world through public-switched telecommunications networks, just like the Iridium system.

After considering all its options, the Inmarsat Council has decided to forgo the LEO concept and develop either a conventional, geostationary system or one that would use intermediate circular orbits. An intermediate circular orbiting system would use up to 15 satellites in an orbit about 6,200 miles above the earth. A decision was expected in early 1994.

The high stakes nature of the MSS market began to emerge in July 1992. At first, Inmarsat said that Project 21 would offer worldwide, pocket-size satellite phone service, and probably not until the end of the decade. But then Inmarsat Director General Olof Lundberg suggested that Project 21 could be extended to other applications; "for instance, entertainment communications centers for your car, personal satellite navigation, position-reporting, and alerting services for your security." Lundberg believes the accumulated worldwide, potential mobile satellite market will be large enough by the year 2000 to support an investment in new satellite systems of about $1 billion.

Writing to Vice President Dan Quayle in his capacity as chairman of the President's Competitiveness Council, Motorola warned that Inmarsat's Project 21 threatened to squeeze Iridium out of the market and suggested that Inmarsat would use its "privileged position" to deter private U.S. LEO companies from obtaining needed investment capital and spectrum. Motorola urged the U.S. government, through its participation in the Inmarsat Convention, and through the FCC, to "restrict the entry of this intergovernmental organization" into the global, hand-held phone market "until the playing field is leveled and there is reasonable opportunity to establish multiple private systems."

Inmarsat developed this briefcase-size satellite terminal for use with its new Inmarsat-M service, the precursor to the international satellite network's proposed global Project 21 system.

Bruce L. Crockett, president and chief operating officer of Comsat, responded to Motorola's charges with his own letter to Quayle. Crockett wrote, "Motorola wants to coerce Inmarsat into an alliance or co-venture for Iridium, or present Inmarsat with the risk of being excluded from a future, satellite hand-held market." Charging Motorola with implying that any mobile satellite service using hand-held phones developed by Inmarsat would not benefit U.S. industry, Crockett pointed out that approximately 25 percent of Inmarsat's revenue is derived from commercial land mobile service, much of which is generated through the use of transportable terminals. At least 15 U.S. manufacturers are already on the market with relatively lightweight, briefcase-size terminals which provide voice, facsimile, and data service on a global basis through the Inmarsat system. These terminals could be reduced to hand-held models by the time Iridium or Project 21 get off the ground.

The Comsat executive called Motorola's charge that Inmarsat is in a "privileged position" a reference to the fact that Inmarsat is an international system—"something Motorola would like to replicate." Crockett wrote that Inmarsat isn't powerful enough to influence market access restrictions in some countries as Motorola seemed to suggest. In fact, the total retail sales value of all Inmarsat partnership services in 1992 was projected at only $600 million, or about 1/24 the size of Motorola.

AMSC has sided with Inmarsat. Its president and chief operating officer, Brian Pemberton, wrote to several high-ranking government telecommunications policy-makers, pointing out that a good relationship with Inmarsat would help coordinate efforts between Inmarsat and AMSC on important domestic MSS spectrum-allocation issues. Pemberton's support of Inmarsat is curious, however, given AMSC's earlier relationship with both Motorola and Inmarsat: On the day Motorola announced its plan to develop Iridium (June 26, 1990), AMSC signed a memoranda of understanding with Motorola, Inmarsat, and Telesat Mobile to jointly explore the potential of Iridium. That agreement has since been dissolved.

While Inmarsat is relying on its signatories to advise the organization on both the technical and economic aspects of Project 21, the signatories are purposely taking their time deciding if they are willing to undertake the financial risk involved in the program. Sweden, for example, which has a 10 percent stake in

Inmarsat, mainly because of its maritime interests, has indicated that it may not be willing to support Project 21. Comsat, on the other hand, with a 25 percent equity position in Inmarsat, may actually insist on a higher level of investment in Project 21 to cover the U.S. MSS market. What is not yet clear is how Inmarsat would operate in the United States. The international satellite consortium is forbidden by the FCC from offering land mobile services in the U.S., where MSS service may only be provided by the AMSC.

Also, in late October 1993, Motorola reminded the FCC that the 1978 congressional act establishing Comsat as a member of Inmarsat does not allow it to offer a satellite-based phone system. The petition also pointed out that Inmarsat had an unfair advantage because it doesn't pay any taxes or duties. "Until the playing field is level," Motorola said in the petition, "we have to just keep on trying to level it."

Virtually forgotten about is a joint venture announced in 1990 by Unisys Corp.'s Defense Systems unit and Colorado-based Energetics Satellite Corp. to build and operate a satellite system offering messaging, position-location, and remote equipment–monitoring services. Under this agreement, Unisys was to design, develop, install, and operate the system's ground stations. Subcontractors like Intraspace Corp. were to produce the satellites and Space Commerce Corp. had agreed to provide launch services using Russian Proton rockets. The proposal is in limbo as Energetics continues to look for funding for the program.

Motorola says it expects to charge $3 a minute for an Iridium phone call. Inmarsat believes it can provide virtually the same service for less than $1 a minute, and Ellipsat says it expects to charge 50 cents a minute. By comparison, a cellular phone call costs from 20 cents to about 95 cents a minute. Market research by TRW indicates that at $3 per minute, only 300,000 subscribers are likely to sign up for service. A lower rate—for example, 30 cents per minute—would generate 12 million subscribers, but none of the proposed systems will have that kind of capacity. TRW says it can make a substantial return on its investment with 2 million subscribers. But those projections could be upset if, as some industry observers have suggested, market conditions force Motorola and its competitors to rent, as well as sell, their phones to their service subscribers.

Motorola isn't alone in its quest for cash. AMSC already has

$346 million to begin full commercial service and claims it needs only another $150 million to $200 million to bring the system to market. But analysts believe AMSC will actually need more funding, perhaps $350 million, while it builds its customer base and begins to pay off its vendors who helped finance the system. TRW continues to raise funds for Odyssey, which is expected to cost $1.3 billion. TRW has said that it would "explore" strategic service partnerships with several telephone companies and others and has issued a private prospectus for Odyssey. Globalstar's original cost estimate of $850 million has jumped to well over $1 billion, and Loral Qualcomm has been looking to telephone operating companies, both U.S. and foreign, and cellular carriers as potential investors. Ellipsat received financial help by forming a strategic alliance with Cairncross Holdings to offer mobile satellite communications in Australia and the surrounding Pacific region.

In an effort to speed up the MSS licensing process, Motorola and Loral Qualcomm have jointly asked the FCC to rule that qualified systems be assigned spectrum as they become operational, with the first system authorized to operate over all the available spectrum. As additional systems become operational, the available spectrum would be divided equally among the operational systems. In addition, the filing asks that licensees be assigned spectrum only as they become operational.

Other essential features of the band-sharing plan proposed by Motorola and Loral Qualcomm are (1) that the FCC allocate the entire 16.5 MHz of spectrum allocated at the 1992 World Administrative Radio Conference (WARC '92) in Madrid in the L-band to MSS and avoid imposing overly restrictive operating conditions in the lower portion of the band; (2) that the FCC permit all qualified applicants to construct their proposed satellite systems with the capacity to operate in the entire allocated frequency bands, in accordance with their applications and a specified milestone schedule; (3) that spectrum assignments be made in a manner that will minimize the potential for interference between systems using different technologies, and between those systems and radio astronomy sites; and (4) that the FCC establish as a legal and policy matter that eligibility for MSS/RDSS licenses in the relevant bandwidths will be limited to nongeostationary MSS systems. The filing also asks the FCC to establish certain financial qualifications and technical standards for all applicants, as well as establish construction and launch milestones for all qualified MSS/RDSS permittees.

In response, the AMSC agreed with the Motorola/Loral Qualcomm provisions covering RDSS, but not the MSS proposals. As the FCC takes up these issues, most disputes are expected to center on when frequencies are assigned—before launch or when the satellites become operational.

Foreign interest in mobile communication satellites is high, but low profile, preferring to monitor the U.S. LEO MSS scene. But it's not dormant. In France, the Centre National d'Etudes Spatiales (CNES) is developing the Taos system, using five small satellites that would offer messaging, position-location, and remote monitoring services. France hopes to launch its first Taos satellites in 1997. The European Space Agency (ESA) also is considering the development of a high-earth-orbit (HEO) system to provide the best possible coverage for European users. In Japan, NEC Corp. is developing small, low-cost satellites.

In Mexico, LeoSat Panamericana, S.A. de C.V. was formed in 1991 by Manuel Villalvazo, who owns Baja Celular-Mexico and is chairman of the board of the Latin American Cellular Association (Alacel), to develop an "affordable" LEO network offering messaging, paging, fax, electronic mail, and position-location services throughout Latin America. LeoSat has been using high-altitude balloon flights to perform frequency interference, communications protocol, and system tests. Following those tests, LeoSat plans to use two LEO satellites to validate its designs in an operational system. Another possible market entrant is Ariadne, a joint Ukrainian-Russian venture, which has proposed a Little LEO network.

Four U.S. Little LEOs have been formally proposed: Leosat Corp., Orbital Communications Corp. (Orbcomm), Satelife, Starsys Global Positioning, and Volunteers in Technical Assistance (VITA), a nonprofit organization that proposed a two-satellite Little LEO network for Third World countries.

Orbcomm has already launched a so-called capabilities demonstration satellite on a Pegasus rocket. The 32-pound package is a precursor to the Orbcomm LEO two-way messaging and data network. Orbcomm plans to begin service in the United States in 1994, launching the first of 2 of its 26-satellite constellation shortly before offering the service commercially. The others are Courier I and Gonets/Small Sat. Orbcomm and Teleglobe Inc. of Canada plan to jointly finance and operate the Orbcomm system. Teleglobe is expected to provide $80 million for the system of up to 26 satellites; Orbital would provide $55 million for the project. Under their agreement, Teleglobe would operate Orbcomm outside the United States under a new company, Orbcomm International Corp.

Starsys plans to transmit and receive very brief, nonvoice radio signals using satellites and ground stations to link customers with a variety of small, low-powered, low-priced portable terminals resembling pagers. Ground processing of Starsys signals will provide the geographic location of transmitting terminals. Applications of Starsys technology could be used for locating stolen vehicles and equipment, monitoring fleet and cargo movements, environmental and utility monitoring and data transfer, and convenient and inexpensive brief, two-way messaging. Starsys, which has already successfully tracked and processed data from an experimental satellite, has signed 15 partners in seven countries and expects to receive its operating license from the FCC by mid-1994 and begin deploying its 24-satellite Little Leo network in 1996.

The FCC adopted proposed operating rules for Little LEOs in early 1993 and hoped to issue operating licenses for the smaller satellite systems by the end of 1993. With their position-location capabilities, Little LEOs like Starsys would like to market their services to the long-haul trucking and other commercial vehicle operators. That could put the Little LEOs in a position to compete with such services as OmniTRACS, a two-way text-messaging and position-location satellite-based communications system already installed in about 40,000 trucks in the U.S.

Developed by QUALCOMM, OmniTRACS leases channels on GTE Spacenets' GStar I satellite. Position location is provided by a Loran C receiver in the truck. In Europe, OmniTRACS is called EutelTRACS because it operates from the Eutelsat satellite network. The system is managed by Alcatel under a joint venture agreement with QUALCOMM and is projected to grow at a fairly good rate as the European trucking industry works to improve its efficiency under the rules and regulations of the single-market European Community. In Japan, QUALCOMM has formed OmniTRACS KK, a partnership with C. Itoh & Co., one of the world's largest trading companies; Nippon Steel Corp.; Clarion, a major consumer electronics company; and Maspro, a supplier of direct broadcast satellite (DBS) receivers.

Increasingly, OmniTRACS is being used by companies other than truckers. Delta Airlines has installed OmniTRACS on vehicles used to transport jet engines. The U.S. Navy and U.S. Coast Guard are using OmniTRACS to monitor the movement of certain types of cargo. QUALCOMM also sees a major market for OmniTRACS in tracking railroad locomotives as they move across the country; there are 24,000 locomotives in the United States alone.

Another start-up, Mobile Datacom Corp., is seeking FCC permission to offer low-cost, satellite-based mobile data communications and positioning services across the U.S. Mobile Datacom plans to use the RDSS to locate and track vehicles, send messages, and transmit data from small vehicle-mounted, or handheld, terminals. The company plans to purchase equipment from Comsat, which will license its RDSS technology to Mobile Datacom. Comsat acquired patent rights to RDSS from Geostar Corp., which went out of business in 1991. But, unlike Geostar, which focused on the highly competitive commercial trucking industry, Mobile Datacom will target government agencies and the aviation and maritime markets. The system would lease capacity on GTE Spacenet's GSTAR 3 and Spacenet 3 satellites. Like the Big and Little LEOs, Mobile Datacom has been looking for financial backing.

There's another proposal that has many satellite experts scratching their heads. Calling Communications Corp., says it will ask the FCC for permission to orbit a constellation of 900 LEO satellites to provide telephone service around the world through existing phone companies. Called the Calling Network, the plan is to launch the satellites into 21 orbital planes, each containing 40 active satellites, with as many as 4 operational spares. The satellites would orbit at an altitude of about 70 km. Calling Communications has been slow to talk about its financial backing, but estimates the cost of orbiting its system at something over $6.5 billion.

Another segment of the MSS market that is only beginning to be tapped is satellite phone service for commercial airlines. So far, only a few airlines offer satellite telephone service, but the new service could be a $100 million industry within a decade.

Inmarsat predicts that 500 planes will carry satellite equipment by 1995, boosting the international consortium's revenues from satellite voice and data services from about $1 million in 1992 to annual revenues of $50 million within 10 years. Skyphone, a London-based consortium owned by Singapore Telecom, Norwegian Telecom, and BT (formerly known as British Telecom), which supplies telephone services to several airlines, believes that 1,000 planes will be equipped to handle satellite calls in five years. Skyphone, along with the Comsat, and ARINC, a U.S. airlines satellite communications system, are part of another consortium, Globalink, which offers satellite telephone service in the U.S. Another consortium, Paris-based Satellite Aircom provides service satellite communications services to Air France and Lufthansa, although Air France is considering testing a ground-based service during 1993. Satellite Aircom's owners include IDB Aeronautical of the United States, OTC Australia, Teleglobe Canada, and the Société Internationale de Télécommunications Aéronautiques, an international airline telecommunications organization.

The service isn't cheap. Skyphone charges the airlines $6.70 per minute, part of which goes to Inmarsat. Singapore Airlines, the first to offer satellite-based in-flight telephone service in 1991, charges its passengers $8.80 a minute on international flights and is averaging about 20 passenger calls per flight, each averaging about 2.5 minutes. Of course, like most competitive services, the cost of in-flight calls is expected to drop. Inmarsat provides its own international air-to-ground phone service, but only for private and general aviation aircraft.

So far, most commercial airlines use ground-based systems, such as GTE Airfone, Inc., which began installing telephones in airplanes in 1984. But In-Flight Phone Corp., formed by John D. "Jack" Goeken, who also founded MCI, the FTD Mercury Network used by florists, and Airfone, has received most of the attention mainly because of Goeken's reputation for successful start-ups. Also, In-Flight was the first to enter the market following an FCC decision in late 1990 to grant new licenses after ruling that Airfone could not enforce any contracts with airlines that it signed before other rivals were licensed. The FCC has also licensed

Claircom Communications, a joint venture of Hughes Network Systems and McCaw Cellular Communications, Inc.; Mobile Telecommunications Technologies Corp. (Mtel); and American Skycell Corp., a West Atlantic City, New Jersey–based start-up. Claircom began passenger telephone service of its AirOne system aboard Alaska Airlines in early 1993.

Unlike Airfone, which uses analog technology similar to the current U.S. cellular network, In-Flight is starting out with a digital system. Passengers on In-Flight-equipped flights will be able to call from one seat to another inside the plane, monitor the status of connecting flights, access real-time stock market quotations, tune in to live news and sports reports and events, and even play movies and video games—all from their seat, using a handset stowed in their armrest and a 4.5-in. by 6-in. screen mounted in the back of the seat in front of them. In-Flight service will be introduced on USAir 757 and American Airlines Md-80 aircraft and are being tested by Continental Airlines. So far, In-Flight users can make calls only while in flight, not actually receive them; however, the company has developed a system that lets passengers receive messages through a special 800 number. Airfone and In-Flight service are much cheaper than the satellite services; both charge $2 a minute. But, unlike Airfone, In-Flight doesn't tack on an additional $2 to initiate each call.

GTE Airfone is responding to In-Flight and others by equipping more than 400 Delta Airlines aircraft with GTE's digital GenStar system and has already tested the new air-to-ground telephone service on selected Delta and United Airlines flights. In operation, the ground stations direct call signals received from the aircraft to the public telephone network using radios manufactured for GTE by Magnavox Government and Industrial Electronics Co. In addition to the digital radios, the GenStar system features new digital handsets and an aircraft PBX to link onboard phones and ground stations. Installed in the seatbacks, the handsets will have a built-in liquid crystal display (LCD) for menus, prompts, and other messages as well as "RJ-11" jacks for connecting personal computers and portable fax machines.

Airfone is also implementing a specially designed "intelligent network" that will enable passengers to receive calls from the ground as well as call other passengers—not just on their own plane, but on other planes as well. On international flights, Airfone can provide links to Inmarsat's aeronautical satellite systems that enable passengers to use any of the system's services when the aircraft is out of range of GTE Airfone's North American ground network. This satellite-based service has been installed on United and other airlines. GenStar also is designed to access the planned European terrestrial network.

Another new Airfone feature is conference calling for business travelers. Available initially on 11 airlines, this service allows passengers to talk simultaneously with up to four ground parties, both in the U.S. and abroad. Domestic conference calls are $5 per minute, plus a $2 setup fee. Thus a 5-minute air-to-ground conference call with four ground parties costs $27. International conference calls on the Airfone Service are $9 per minute, plus a $4 setup fee.

Already, equipment suppliers are lining up to sell airborne satellite systems to commercial airlines. Key players on the equipment side are reporting orders for 1994 well ahead of anything they experienced in recent years. In most cases, these companies have established teams to offer airlines packaged systems. They include Rockwell International's Collins Air Transport Division/Ball Aerospace Systems Group; Honeywell Air Transport Systems Division/Racal Avionics, Ltd.; and Bendix-King's Air Transport Avionics Division/Dassault Electronique. Others, including Canadian Marconi, a major antenna designer and manufacturer; E-Systems and MIL-COM Electronics Corp., both U.S. satellite communications systems manufacturers; and Toshiba of Japan, a potentially formidable competitor in satellite ground stations, are waiting in the wings. Considering the high-performance requirements placed on equipment by satellite operators and the airlines, it is unlikely that this segment of the communications equipment industry will become overcrowded.

Mobile communications satellites are extremely complex and so is operating them as a business. Hundreds of millions of dollars have been spent by some very smart people in anticipation of LEO and MEO satellites becoming a major market. It is still possible that MSS, which sounded very good just a year or two ago, may turn out to be too risky a concept except for certain well-defined niche markets. It is becoming increasingly difficult, for example, to distinguish between MSS and rapidly emerging digital cellular and personal communications services. If cellular technologies can provide low-cost, worldwide mobile and fixed telephone services, how will that impact MSS?

One big hurdle is the FCC's final ruling on LEO satellites, which was scheduled to be announced in October 1993. The rules were supposed to be based on the negotiations between the applicants, concluded earlier in the year, but there was little agreement on key issues, such as spectrum sharing. With so much material to wade through, the commission may be forced to delay its rule making. How will the new procedure for allocating spectrum take into account the technical viability of totally new communications systems? How much will a license cost? How will small companies fare under the new spectrum auctioning process? Will auctions shorten or lengthen the licensing process? (If foreign experience is any guide, auctioned licensing takes longer.) Lots of questions, which only Congress and the FCC can answer.

So far, the FCC has granted proposed geostationary and LEO MSS frequencies in the 1610–1626.5-MHz and 2483.5–2500-MHz ranges. It isn't clear how the FCC will divide the bandwidths among the MSS licensees. The commission proposes a spectrum sharing plan that would accommodate up to five players; however, most industry analysts believe the market can only support three systems.

MSS is a very-high-risk, high-stakes business, one that has already witnessed the death of several satellite services, including Comsat DBS, Geostar, and Federal Express FaxSat. If the LEOs aren't very careful, MSS may prove to be no different.

Chapter 5

WLANs and WPABXs

No More All Alone (by the Telephone)

The general scenario goes something like this: You have been reorganizing your office at least once a year, moving people around, adding staff, buying more computers. Now you're doing it almost twice a year. It's unavoidable because you're growing and your priorities are changing. But every time you move, add to, or otherwise change your offices, it costs thousands of dollars to recable your computers.

There is an alternative, however: a wireless local-area network (WLAN), which can significantly reduce moving costs, particularly in older buildings where asbestos virtually requires a WLAN. Wireless LANs are also being installed as a backup to existing wired systems in disaster recovery situations or as new LAN nodes that may ultimately be wired but that need to be "connected" immediately.

Wired local-area networks already represent one of the two fastest-growing segments of the computer industry. (The other is mobile computing.) Each year, about half of all companies relocate at least part of their operations. It can cost as much as $2,000 to move an employee. If a company moves 1,000 workers a year, it's going to cost $2 million. A WLAN terminal costs less than $500, or $500,000 for a comparable move. WLANs also remove the long-standing limitations to adding to a wired LAN—the need to physically install it, the need to manage it, and the need to reconfigure it. To WLAN vendors, it's an opportunity to replace "the last 100 feet" of wiring to office desks and equipment.

WLANs already fit into factory or retail environments. Younkers Inc., a regional chain of department stores in Iowa, has installed 2,500 point-of-sale (POS) terminals based on NCR Corp.'s wireless WaveLAN system, allowing employees to access central computer-based information from any store or remote site. California Microwave has delivered several hundred of its RadioLink WLANs to Britain's Marks and Spencer department stores to interconnect the stores' cash registers. American Airlines' SABRE Travel Information Network installs WLANs in travel agencies. SABRE is leasing NCR's WaveLAN to travel agents who want to use it during a move, or as a permanent replacement to their wired LANs.

The initial acceptance of WLANs has been slow. Several major independent market research organizations disagree on the current level of WLAN sales. Estimates for 1993 range from about

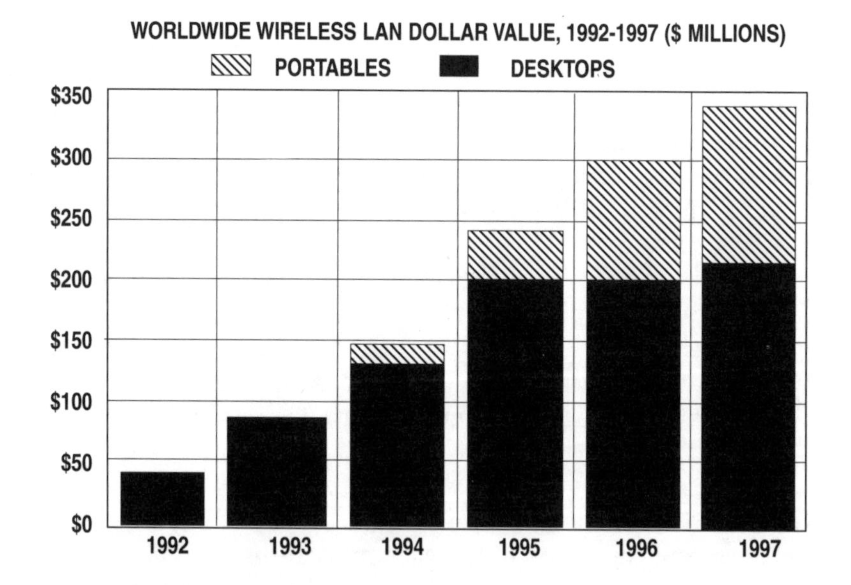

Source: International Data Corp.

Market researcher International Data Corp. expects the worldwide wireless local-area network market to top $300 million beginning in 1997. By that time, almost half of the value of the market will be in portable computer applications.

$100 million to almost $250 million. But they all agree that it's going to be a big market.

Market segmentation may also be confusing: Some WLANs require a license to operate, some don't. Data rates, distances covered by different technologies, and distribution systems also differ. Proxim, for example, uses two-step distribution for its original equipment manufacturer (OEM) board-level products. Windata uses a multitiered distribution channel consisting of network integrators, OEMs, and system integrators. Motorola has broadened its distribution base by signing IBM's Customized Operational Services organization to resell Motorola's Altair WLAN products. Ungermann-Bass Inc., a LAN specialist, not only resells the Altair line, but has agreed to help Motorola develop new WLAN products. It's also unlikely that LAN users will tear out their existing networks for a wireless LAN. More likely, they will begin to install WLANs as adjunct systems, when they are needed and where they're cost-effective. And, unlike most industries, there is no clear WLAN market leader.

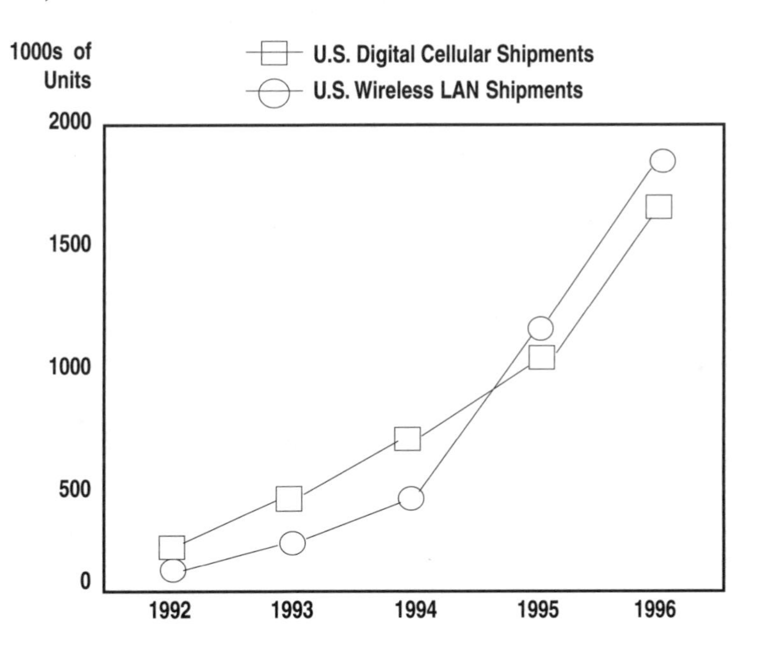

Source: BIS Strategic Decisions

U.S. WLAN shipments are expected to closely track U.S. digital cellular sales. Both are rapidly emerging wireless markets.

If there's a show stopper for WLAN, it's standards; there aren't any, at least, not yet. With incompatible WLAN products starting to enter the market, that will have to change. Without standards, users are worried about, as one industry executive put it, "betting on the wrong horse." They don't want a rerun of the Betamax/VHS standards fiasco.

Two industry groups have been working on the standards issue. The Institute of Electrical and Electronics Engineers (IEEE), the largest technical society in the world with more than 300,000 members, has established a subcommittee—IEEE 802.11—to produce a 1- to 10-megabit per second (Mbps) wireless LAN specification covering everything from wire augmentation and ad hoc networking to factory automation. The IEEE group has been meeting regularly for more than two years, but progress has been slow. The group has announced its support of a unified medium-access control (MAC) "foundation" for WLANs. The MAC represents a basic framework upon which enhancements and refinements will be added in future 802.11 proceedings. To accommodate this protocol, all current equipment vendors will be required to modify their existing products to ensure compliance. The new standard will support more than 1,000 nodes operating within several hundred meters of each other at transmit speeds up to 20 Mbps (current wireless products support up to 2 Mbps). However, there is more work to be done on both the MAC and physical layers of the WLAN standard. In fact, a draft standard for a fully functional WLAN isn't likely until at least the end of 1994.

The other organization is WINForum, a relatively new, but aggressive group made up mainly of Silicon Valley–based communication, computer, and semiconductor companies. WINForum was born shortly after Apple Computer filed a petition with the FCC asking for the allocation of 40 MHz of spectrum in the 1850–1990-MHz band for a new local-area data network Apple calls Data-PCS.

The Data-PCS proposal evolved out of three major trends that Apple's product planners had identified among personal computer users. The two most important are connectivity and portability. Put those two features together, and you have a wireless data network. The third trend has more to do with the kind of computing people are doing today, that is, using various media—what Apple calls compound documents—to produce images and graphics. In the future, even the smallest computers will be able to handle video clips and sound annotation. These applications will require higher data rates. Data-PCS is an attempt to recognize all these trends and anticipate the networking requirements that go beyond what is possible today. Ideally, in Apple's corporate mind, Data-PCS would become an open, international standard, one that everyone can use.

The Data-PCS petition itself proposed only general objectives and a regulatory framework for meeting them; it did not offer technical specifications to be incorporated in FCC rules for Data-PCS. In fact, Apple even suggested in its petition that "there should be thorough dialogue within the industry and between the commission and the industry" to refine the definition of Data-PCS. With few details about what Data-PCS should actually look like or how it might function, companies and trade groups began to respond to Apple's petition with some very specific ideas of their own, most of which had little to do with Apple's concept. The FCC became confused about what the industry really wanted and said so. The result was the formation of WINForum—to

The Freeport ™ Wireless LAN System from Windata Inc. is based on spread-spectrum radio technology and features Ethernet® compatibility.

present a common front on wireless issues of interest to both computer and communication equipment manufacturers. Within a very short period of time, WINForum began promoting a generic concept called User-PCS in which user-owned wireless personal communications devices would be available without a license or a per-call charge—an important feature to computer manufacturers who don't want to license their products, especially portable laptop and notebook computers.

Despite the wide use of spread spectrum (close to two-thirds of all current WLAN installations use this technology), it has some technical limitations. For one thing, it is bandwidth-limited. Even with multiple frequencies, the total bandwidth available to spread-spectrum systems is inadequate to support more than about 1 Mbps, only a tenth of the minimum 10 Mbps required to be compatible with current wired LAN data rates. Boosting the data rate won't help much because it tends to reduce the range in spread-spectrum systems. Adding base stations would cover this problem, but that means adding cost. This could also cause interference problems with other spread-spectrum users, namely, cellular telephones, FM radio, and the upper end of the UHF television band.

Key players in the three WLAN technologies include

Spread-Spectrum: NCR, California Microwave, O'Neill Communications, Windata, Symbol Technologies, Proxim, Digital Ocean, Telesystems SLW, Xicom, and Dayna Communications.

Narrowband Microwave: Motorola

Infrared: Photonics, InfraLAN Technologies, and Radiance Communications.

Ing. C. Olivetti & Co. S.p.A., the Italian computer giant, has announced a Net3 WLAN system based on the Digital European Cordless Telephone (DECT) standard. Olivetti believes working with an established standard is important; the company's technical staff participated in the development of DECT.

Several start-up companies are either testing WLANs or are just coming to market with WLANs or wireless connectivity products that fit into WLANs. Key among them are ADC Kentrox, Armatek, Spectrix, Tellular, and VeriFone.

Motorola introduced its Wireless In-building Network (WIN) in October 1990. It is the only narrowband microwave WLAN on the market and the only WLAN transmission scheme that requires FCC licensing. Called Altair, the system operates at 18 GHz, a frequency set aside by the FCC in April 1990 to be used inside buildings. Unlike other WLANs, Altair is designed to replace, rather than supplement, existing wired LANs, such as Ethernet and Token Ring networks. This makes it transparent to existing communications standards and protocols. However, Altair is a much more expensive system than the other WLANs.

The major advantage of the 18-GHz band is its propagation characteristics for a microcellular network. The same frequencies can be reused by another system within 120 feet or so, or on the other side of a dense barrier, such as a cement floor. Also, 18 GHz is a high enough frequency that office and factory equipment aren't likely to interfere with it. Another important feature of the system is an antenna that uses a signal sampling scheme to automatically and continuously select the best signal for each data transmission. Motorola claims the antenna addresses the data security issue by making the network signal stream virtually incoherent to an outside receiver.

Since its introduction, Motorola has announced a number of product upgrades, including an 80 percent improvement in the WIN's throughput performance. Now designated Altair II, the system can support up to 50 user modules. A separate product, the Altair Vista, is designed for building-to-building communications in a campus environment.

Infrared (IR) systems use a part of the electromagnetic spectrum just below visible light as a transmission medium. IR technology is well suited to very high data rates; systems having data rates of 100 Mbits/sec have been developed. However, being a light medium, IR communications are limited to a straight, or line-of-sight, path (like your TV channel changer, for example) and can be blocked by opaque objects. Systems that use diffused sources and that rely on reflections off walls and ceilings have been built, but none have succeeded in multiple-room environments without the use of extensive repeater networks. Also, IR transmitters and receivers are easily interrupted by moving objects, such as people. Still, some industry analysts believe that products like Apple Computer's notebook-size "personal digital assistant," or PDA, give credibility to IR technology. Obviously, Apple agrees; it invested $5 million of Photonics' $6.5 million start-up funding and is believed to own an 18 percent stake in the company. (Not to be outdone, Digital Ocean has introduced a spread-spectrum–based WLAN for Apple's Macintosh PowerBook laptops and LocalTalk-compatible devices. Called Grouper, the unit can transmit data between PCs at distances up to 500 feet.)

Although initially developed as a cost-effective option for wired LANs, new developments have opened up WLAN applications to portable computer users, offering peer-to-peer communications with other portables for such applications as E-mail, printer sharing, modem sharing, and file sharing.

A key enabling technology for portable computer users is the standard developed by the Personal Computer Memory Card International Association (PCMCIA). PCMCIA slots in portable computers allow users to implement peripherals typically found in desktop computers, such as LAN adapters, in their portable

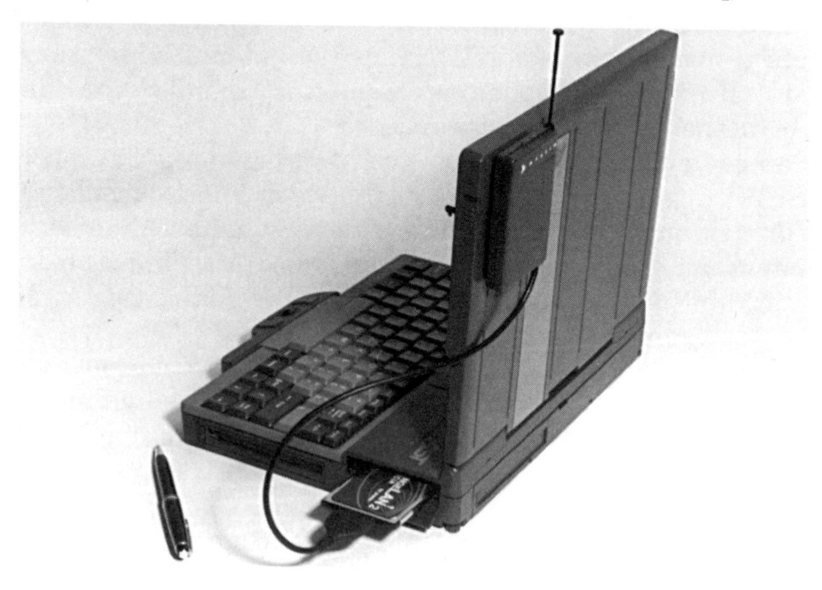

Proxim's RangeLAN2 WLAN, based on frequency-hopping spread-spectrum technology, incorporates many of the features required in the IEEE 802.11 wireless local-area network foundation protocol.

models. The PCMCIA standard specifies form factor, a standard connector configuration, and standard interface protocols for portable peripherals and has become increasingly popular in new laptop, notebook, and palmtop computers.

Proxim has begun shipping a PCMCIA WLAN adapter designed to fit a 5-mm-thick PCMCIA card socket. This size card is usually used for modems, LAN adapters, and host communications cards. The Proxim RangeLAN/PCMCIA allows mobile computer users to communicate with existing wired client/server networks or to set up a peer-to-peer LAN between mobile PCs. So far, Proxim has demonstrated its RangeLAN/PCMCIA using Microsoft Windows for Workgroups and Microsoft Windows for Pen Computing. NCR has announced a similar product for its WaveLAN WLAN. Many other WLAN/PCMCIA products are expected to be introduced in the next few years, some from companies not mentioned here. Typical of the rate of development of computer and communications technologies, new WLAN products will continue to offer more advanced features and probably will cost less than they do today.

Wireless private automatic branch exchanges (WPABXs) are used primarily for voice communications, although some companies have been using them successfully as a low-data-rate WLAN system. WPABXs are being pitched as the way to reduce wiring costs and virtually eliminate unanswered calls by freeing workers to roam around the office or plant with a cordless phone that is linked to the company's WPABX. This may be viewed as more of a curse than a blessing, but studies have shown that some 60 percent of business calls do not reach the intended party on the first try. Voice mail has alleviated this problem to some extent, but wireless phone access in the office could potentially regain much of the lost productivity and lost business opportunities that result from missed phone calls.

Although an all-wireless PBX system is not expected for the next several years, wireless adjunct systems are now available that link to a PBX. Future plans call for specialized circuits (called T1 links) that transmit digital signals over wired telephone networks between the adjunct and main PBX.

Wireless PBX adjuncts usually consist of a control server and strategically placed transmitters/receivers, or base units, including antennas. Pocket telephones with a range of about 100 meters communicate with the base station, which is usually hidden away in a closet. As the user roams the facility, calls are handed off among the base stations, just as they are in a cellular network.

The wireless PBX market is off to a slow start for many of the same reasons WLANs are experiencing slow market acceptance—MIS and telecom managers are trying to figure out how to implement wireless technologies in their office and factory systems. They're concerned about spectrum-allocation and standards issues, and proposals before the FCC that on-premises wireless systems carry airtime charges or frequency use fees. Clearly, PBX users would be reluctant to pay for airtime when using their own system. And they haven't yet seen enough WLAN or wireless PBX installations to give the technology the credibility to commit to a full-blown, in-house wireless network.

Results from a study of actual users of in-building wireless communication systems (WCS) and a nationwide survey of more than 400 business organizations indicate a strong need for these systems. But WCS users want a lower-priced product than is currently available, and they have very specific performance requirements. The study, which was conducted by Alexander Resources, also found that if customer requirements can be satisfied, WCS will replace or augment 26 percent of all existing wired PBX or Centrex (so named because it uses central office switching equipment) and key telephone systems by the year 2000.

Initially, Alexander Resources believes that customers will purchase and use WCS to improve employee productivity and communications. Later, as WCS price and performance reach that of a wired telephone system, WCS purchase justification will be made more on cost. The study also found that potential in-building WCS users don't believe that cellular or cordless telephones will meet their needs; they require performance and capabilities similar to those of a wired desk telephone.

The results of a survey by the North American Telecommunications Association (NATA), which represents the PBX/PABX community, are more favorable for the market, portraying the potential benefits of wireless business systems "as great or greater than those promised by any other segment of the emerging personal communications marketplace." (NATA's interest in wireless systems has grown as its members' traditional markets have shrunk—the value of the PBX market decreased steadily from

$3.3 billion in 1985 to $2 billion in 1991. This lackluster performance is likely to improve only slightly through the early 1990s.) NATA's survey found that early participants in the wireless business telephone market are using the same analog cordless telephone technology that is so popular in the residential market today. But since existing cordless residential phone designs don't allow users to move beyond the range of their own base station transmitters/receivers, manufacturers are developing wireless premise equipment, such as WPBXs, that incorporate the capabilities of cellular telephone systems with such features as handoff, roaming, and automatic call delivery. Adding cells would enhance the functionality of wireless communications in an office by allowing users to initiate, answer, and continue telephone conversations when they move beyond the range of their "home" base stations.

To satisfy the potential demand for wireless communications in the workplace, NATA has urged the FCC to assign frequencies to wireless in-building applications that would supersede existing cordless telephone frequencies for unlicensed use covered under the FCC's Part 15 rules. Because the FCC requires that cordless telephones and other wireless devices registered under Part 15 not interfere with FCC-licensed services, there may be some inherent risk for anyone who purchases a wireless PABX. This risk, NATA believes, is deterring many equipment manufacturers from advancing the development of wireless business premise systems.

To date, most of the action has been in Europe because the region has well-established standards. In Japan, the Telecommunication Technology Council, under the Ministry of Post, Telegraph and Telecommunications, has established a committee to develop a WLAN access method and protocol. In the United States three start-ups, SpectraLink Corp., Omnipoint Corp., and Rose Communications Inc., are building systems that operate in the ISM frequency band. AT&T offers a cordless phone for its Merlin key systems for commercial use, but has not yet introduced this capability on other advanced systems, such as its Definity Generic 1 PBX. Stanford Telecommunications, Inc., is developing a next-generation digital cordless telephone system for PBX and PCS applications for a consortium of five Taiwan companies organized by Taiwan's Industrial Technology Research Institute.

SpectraLink's Pocket Communications System uses small spread-spectrum–based radio receivers called remote cell units (RCUs) to relay calls between 6-ounce pocket telephones and the PBX. RCUs may be placed wherever wireless service is needed. As the user moves throughout the coverage areas, conversations are automatically "handed off" from RCU to RCU. The system provides up to 2 million square feet of coverage and can support more than 700 users, and systems can be linked when additional capacity is needed. SpectraLink is also introducing an outdoor RCU to accommodate calls between buildings and parking lots, or for use in campus environments. SpectraLink already has distribution agreements with Ameritech and Fujitsu Business Communication Systems for its WPBX adjunct product.

Southwestern Bell Personal Communications is using SpectraLink equipment in experimental trials at the University of Texas M. D. Anderson Cancer Center and St. Luke's Episcopal Hospital. Employees at these sites will be assigned one telephone number and can make and receive calls on a pocket phone anywhere in the vicinity of the network. In this test, users can screen calls or route calls anywhere they choose.

The hospitals provide a natural environment for the trials. At M. D. Anderson and St. Luke's, like most hospitals, doctors are paged electronically or audibly over a public address system. The doctors then must go to a phone to find out why they were paged. With the Southwestern Bell test system, doctors, nurses, and hospital administrators have instant, private, two-way voice communications.

At the University of Illinois at Chicago, Ameritech, the Chicago-based parent of Bell companies serving Illinois, Indiana, Michigan, Ohio, and Wisconsin, and other information-related subsidiaries, has combined the features of Centrex and lightweight, wireless phones to link employees at the UIC Pavilion, a special events center, into the university telephone network. Base stations have been placed strategically throughout the building and campus. They receive signals from, and route signals to, the portable telephones. Calls are routed through the PBX/Centrex to a controller, which manages system access and routes calls to the base stations. Employees can place and receive calls anywhere in the Pavilion. Ameritech is making the system available to customers throughout its five-state region.

More than 150 BellSouth employees are using portable, pocket-size telephone handsets in office buildings and on the streets of midtown Atlanta as BellSouth Enterprises expands as part of the regional telephone company's ongoing and expanding PCS trials.

Trial participants were issued 6-ounce SpectraLink 2000 handsets. With the technical aspects of the trials completed, BellSouth is considering the qualitative issues. How are customers using the service? How long do they talk? How convenient is the service? Where are people using it? The findings from this test will be used along with the results from other trials to help BellSouth determine what customers want and expect from this type of personal communication service.

GTE Corp. has put together a wide-area wireless Centrex and PBX service to five businesses in two GTE markets—Nashville, Tennessee, and Durham, North Carolina—as part of its PCS research effort. Called Tele-Go Business Service, the 350-handset business market trial marked the first time that tariffed Centrex services have been linked to wide-area wireless services. This allows business users to take advantage of all PBX and Centrex services from any location throughout an estimated 3,500-square-mile PCS coverage area. GTE customers have coverage within their office or "campus area" for a flat monthly rate and can continue to use the service when away from this area for a nominal per-minute "premium area" charge. It provides four- or five-digit dialing and PBX/Centrex features such as conference calling, call transfer, and call waiting/hold throughout the extended PCS coverage area. It also provides dual-ringing features of the desk set and the portable handset. This dual-ringing capability is provided through a prototype cellular mobility node, developed and manufactured by Northern Telecom.

The purpose of the trial was to determine how a wireless environment changes the way businesses and their employees function. By analyzing efficiencies and benefits of mobility in the workplace, GTE and the other trial participants hope to better understand the needs of the market.

What NATA and their PBX and other telecommunications equipment supplier-members would like to see is action by the FCC that would alleviate the potential for interference of FCC-licensed services and relieve the spectrum bottleneck that is preventing the wireless premises systems market from reaching its full potential. To help push things along, NATA is urging the FCC to allocate frequencies for wireless premises systems in the vicinity of, but not overlapping, the frequencies allocated for other personal communications services.

One possibility would be for the FCC to adopt an open-entry approach rather than a restrictive licensing scheme for the wireless premises system marketplace—even if a restrictive licensing scheme is adopted for new off-premises systems. This would be consistent with the commission's long-standing premises equipment market entry policies. The open-entry approach would allow the on-premises market to evolve naturally, so that various technologies can be tested on a relatively small scale under actual market conditions. Most business users, for example, would be able to proceed cautiously to a wireless environment by adding a few wireless units to existing PBX or key systems, without risking an immediate jump to a fully wireless office.

Technically, NATA admits that the open-entry approach would require some up-front effort to define the spectrum and power boundaries for premises systems and a workable type-certification or quasi-licensing scheme that can accommodate a large number of diverse products. To foster market diversity, PABX/key systems vendors are pushing the FCC to prevent what they perceive is an anticompetitive and restrictive licensing scheme adopted in the cellular service market. If cellular carriers participate in the premises system market, NATA members believe they should be required to do so through separate affiliates. In addition, they want the FCC to ensure nondiscriminatory access to public networks operated by common carriers, whether they are wired or wireless. Specifically, they want to prohibit the bundling of wireless premises offerings with either wireless or wireline carriers' public network services.

As the WLAN/WPABX markets begin to take off, there will very likely be new entrants, offering fairly unique wireless connectivity products. Some industry observers, for example, believe that AT&T, which acquired NCR in 1991, might add voice capability to NCR's WaveLAN line. IBM may also enter the WLAN arena at some point. ROLM has been working with Siemens, its parent company, and others on a wireless PBX system that could be introduced at any time. Software companies such as Microsoft are also developing features designed to incorporate wireless connections in their products.

Outside the U.S., several companies are building 2.4-GHz spread-spectrum systems because of the potential international market opportunities in Europe and Japan at that frequency.

Ericsson has announced a wireless version of its MD 110, called the DCT 900 PBX, that supports cordless phones. It operates at 800–900 MHz, and it is based on CT-3, or third-generation digital cordless telephone technology, which Ericsson pioneered. It supports paging and hand-held data device services as well as voice communications. Meanwhile, Peacock AG, a $300 million personal computer vendor based in Haaren, Germany, is working with Motorola to produce prototypes of two cordless office systems; Motorola is contributing the radio technology, Peacock the switching and call management hardware and software.

Northern Telecom has introduced its Companion wireless PBX/Centrex system in Hong Kong, where several thousand CT-2 phones are already in use, and has signed distribution agreements for Companion in Europe with PTT Telecom in Holland, where it will be marketed as Vox Cordless Companion, and with Televerket, the national telecommunications operator in Sweden, which plans to make it available as part of its Cosmos 2000 product series.

Companion is based on the common air interface (CAI), a specification developed by the U.K.'s Department of Trade and Industry that permits CT-2 equipment made by different manufacturers to communicate with each other. One of Companion's features is that it allows incoming calls to ring on a desk phone and portable handset at the same time. If you're receiving a call and you are away from your desk, you can use the portable unit, return to the office, and transfer the call to the desk phone by pressing the speakerphone key and closing the flap on the handset. Another feature is three-way conference calling. Initially, Companion will use a CT-2 handset developed by Motorola.

Most of the U.S. wireless PBX suppliers are start-ups, such as SpectraLink, whose 2000 Pocket Communications System is capable of supporting 2,000 lines and 400 simultaneous calls.

But no matter who manufactures the product, or what technology is used, what really counts are price and performance.

Chapter 6

International Cellular/PCS Markets

International markets, including Europe, the Asia-Pacific region, eastern Europe, and the former Soviet republics are turning out to be every bit as exciting as the United States. In fact, Europe and several regions of the Far East have already jumped ahead of the U.S. in both digital cellular and PCS.

The most significant development in Europe is the Global System for Mobile Communications, or GSM. As Europe's new digital cellular standard, it is expected to push the cellular penetration rate well beyond today's highs of 6 percent in the strongest European markets. By the end of 1993, virtually every European country, and many eastern European states, will have at least one cellular system. As a result, most of Europe's post, telephone, and telecommunications (PTT) authorities view GSM as a threat to their main source of revenue—the plain old, wired, telephone service.

Initially known as Groupe Speciale Mobile, GSM evolved out of the rapid growth of conventional analog cellular networks in Europe and concern over the need for a system for the future with much greater capacity than analog could offer. In 1982, the Conference of European Post and Telecommunications (CEPT) formed a special working group to develop GSM. By 1987, 18 European nations had committed to the technology by signing the GSM Memorandum of Understanding. In 1989, the effort to develop a GSM standard was shifted to the European Telecommunications Standards Institute (ETSI), and plans were put in place to begin GSM service in 1991. Things didn't quite work out as planned; the entire effort was slowed by almost a year when GSM equipment wasn't widely available and technical problems began to emerge; some were software related, but the big one was incompatible equipment from different manufacturers. Roaming became an issue when GSM operators tried to work out agreements on how to charge subscribers who use their GSM phones outside their home territories. Most of these problems have been solved, and GSM appears to be on its way to becoming the world's most widely used digital cellular standard . . . at least, for awhile.

So far, at least 36 countries have committed to GSM: Australia, Austria, Bahrain, Belgium, Cameroon, China, Czechoslovakia, Denmark, Finland, France, Germany, Greece, Hong Kong, Hungary, India, Italy, Kuwait, Luxembourg, Malaysia, New Zealand, Norway, Poland, Portugal, Qatar, Romania, Russia, Singapore, South Africa, Spain, Sweden, Switzerland, Taiwan, Thailand, Turkey, United Arab Emirates, and the United Kingdom. Most African nations are also expected to adopt GSM.

But GSM, which operates in the 849–890-MHz, 935–960-MHz, and 1.7–1.8-GHz frequency bands, still competes with analog cellular throughout most of Europe. In fact, Italy and Spain have only recently begun analog cellular service and may want to continue with analog for as long as their assigned spectrum can keep up with the demand for service.

GSM is emerging in two phases. Initially, GSM will handle basic voice service and some emergency calling features. The second phase, scheduled to begin in 1994, will add call waiting, caller information services, and improvements in subscriber identity module (SIM) cards, which contain a microchip with information on the caller. By inserting a SIM card in a GSM phone, the caller can gain access to the GSM network, and be billed for the call, even if it isn't his phone. SIM cards can also store abbreviated phone numbers for speed dialing.

From the users' point of view, the obvious difference between GSM and the digital cellular systems now emerging in the U.S. is that, in Europe, digital cellular phones operate only digitally, in the GSM mode. In the U.S., cellular carriers are offering dual-mode (analog/digital) phones, and U.S. subscribers have the option of switching between the two modes. Digital-only cellular phones aren't expected to be marketed in the U.S. for at least two to three years.

Market studies sponsored by the European Community Commission (ECC) indicate a potential subscriber base of some 10 million users of GSM phones by the end of the century. A study by London-based CIT Research projects 18.5 million GSM subscribers by 2002. By then, CIT says annual sales of GSM phones could reach 6.2 million. InfoCorp Europe, a Paris-based market research group, expects GSM to begin to surpass analog networks within three years and build to 19 million GSM subscribers in Europe by the year 2000.

Another major development is the emergence of Digital European Cordless Telecommunications (DECT). Most analysts expect analog cordless phones to be almost totally phased out in Europe by 1997, but they don't agree on how the market will be split between CT-2 and DECT phones. Initially, the big market for digital cordless will be in PBX systems, but cordless may eventually account for two-thirds of the residential and telepoint markets. DECT supports both voice and data with encryption, and seamless handoff between cells, using spectrum already allocated throughout Europe in the 1880–1900-MHz band.

DECT has the support of ETSI, ECC, and five major European telecommunications companies. Alcatel, Ericsson, Nokia, Philips, and Siemens have all announced plans to show DECT products and concepts with the expectation that DECT will begin to take off as a market in early 1994 in several European countries. Taiwan also plans to make its first major move into the digital wireless communications market in Europe with a DECT-based system: Hsinchu-based Computer and Communications Laboratory has organized a consortium to develop DECT-based products and pilot projects to produce DECT-based handsets, and PBXs were expected to get underway in early 1994.

Another new service, which could emerge as Europe's PCS or PCN (personal communications network is the preferred terminology in Europe), is DCS 1800, or digital communication service at 1800 MHz. An extension of GSM, DCS 1800 is being standardized by ETSI. But it is still not clear how the basic GSM service will be affected by the emergence of DCS 1800, particularly since the two systems will overlap in most regions, with the GSM network and handsets possibly serving as the backbone of the PCN microcell network. This technique would give GSM service providers an easier and much less costly path to PCNs, even though PCNs require more transmitters for their short-range signals.

The United Kingdom represents the largest wireless market in Europe with slightly more than 1.3 million cellular subscribers at the end of 1992. It is also the most confused market. The British government licensed four companies to provide CT-2 telepoint service shortly after the publication of the landmark "Phones on the Move" by the Department of Trade and Industry, but they couldn't handle the high operations costs, the competition and, least of all, the lack of interest. All four soon disappeared. Today, the only company offering telepoint service in the U.K. is Hutchison Personal Communications Ltd., which wasn't even one of the original licensees.

Hutchison has invested a reported $100 million to build more than 8,000 base stations in train stations and along major highways in the U.K. But the service, known as Rabbit, never really caught on. (The joke in the U.K. is that there are more telepoint base stations than subscribers, and, in fact, only a few thousand people have subscribed to the service.) Hutchison has tried cutting its losses by selling CT-2 phones directly rather than through retail outlets and cut the price of the phones from $184 in mid-1992 in the U.K. to as low as $69 in early 1993.

Shortly after it licensed the telepoint operators, the British government issued personal communications network licenses to three companies—Mercury Personal Communications, Unitel, and Microtel—that said they planned to offer nationwide service. Mercury was initially a joint venture between Mercury, a subsidiary of Cable & Wireless, and US WEST. Unitel decided to abandon its PCN plans shortly after receiving its license and merged with Microtel, which is now owned by Hutchison. (Microtel had been owned by Pacific Telesis and Millicom, both U.S. companies, and Matra, a French company, but they all dropped out of the venture.) The Microtel-Unitel partnership still owns a telepoint license, but it's not clear what they're going to do with it.

One of the problems in getting a wireless personal communications network off the ground in the U.K., according to "British PCN Policy Pitfalls: Implications and Lessons for the U.S.," a report commissioned by the Cellular Telecommunications Industry Association (CTIA) and written by Alan Pearce, president of Information Age Economics, Inc., was the high cost of building three separate PCN infrastructures—about $1 billion each. Another was the way in which the British government doled out spectrum for the PCNs. As Pearce points out in his study, the three PCN licenses were a huge spectrum giveaway. The government, he says, assigned 150 MHz of spectrum at 1.8 GHz (1800 MHz) and said that the licensees were entitled, once their networks had been built, to 50 MHz each. That caused an immediate outcry of unfair competition, since cellular carriers in the United Kingdom only have 30 MHz of spectrum.

Government licensing rules also call for universal PCN service, but licensees say they plan to launch limited service focusing only on major metropolitan areas. In other words, says Pearce, universal PCN will have to wait until PCN proves itself in the marketplace. The dilemma here, according to the Pearce report, is that the PCNs can't have the full 50 MHz of spectrum until they offer universal service, and they can't afford to build a universal infrastructure until they prove that PCN service is in demand in the marketplace. Meanwhile, the PCN licensees are operating on 10 MHz of spectrum, temporarily quieting the two cellular carriers, Vodafone and Cellnet.

The loudest response to the Pearce report came from US WEST, a founding member of the CTIA and a partner with Cable & Wireless Plc in the U.K. US WEST publicly disassociated itself from the CTIA-sponsored study. In a letter to FCC Interim Chairman James Quello, US WEST said the CTIA is using the Pearce document to support its campaign in Congress to limit the potential of PCS. "The CTIA's recommendation that the public would best be served if the United States were to authorize five PCN licensees [per market] is not based upon the U.K. experience and represents no more than an attempt by the cellular industry to avoid viable competition," US WEST said in its letter to Quello. US WEST also said it believes that the record of the U.K.'s PCN experience bears out the wisdom of certain policies, including awarding no more than three PCS licenses and allowing these licensees "latitude in meeting customer needs," using large regions for licensing PCS, and providing each licensee with access to a substantial block of spectrum to reduce the cost of providing service and to stimulate a mass market.

In response, the CTIA called the US WEST assertions that it

was using the report to gain influence in Congress "spurious," noting in its own letter to the FCC that the "CTIA and its members have been active, vocal advocates in the commission's PCS proceeding, urging a broad definition of PCS and open entry for as many competitors as possible."

Meanwhile, the telecommunications industry is keeping a close eye on the Mercury One-2-One PCN service, which is confined to London. Despite call charges that are up to 20 percent lower than cellular service in the U.K., service restrictions and reports of minor technical problems may make it difficult to wean cellular users over to the newer service.

With little acceptance of PCNs, the two major cellular carriers in the U.K., Cellnet, a subsidiary of British Telecom, and the Vodafone Group, which is owned by Racal Electronics, have restructured their tariffs to be more competitive with PCN services. Vodafone launched a regional GSM service in December 1991, but it met with little initial success. The service was essentially relaunched in early 1993, targeted at business users. Since then, Vodafone, the larger of the two U.K. cellular operators, has signed roaming agreements with SIP, the Italian telecommunications concern, and Société Française de Radiotéléphone (SFR), one of the French cellular system operators. Vodafone has also introduced a "short message" text transmission service using GSM protocols and plans a low-cost derivative of GSM, called the Micro Cellular Network (MCN), that would compete with PCNs. If it's reasonably successful, Vodafone says it will consider introducing MCN in other markets, possibly starting in Australia. Cellnet was expected to launch a GSM cellular service in 1994, starting regionally before going national. However, the initial high cost of GSM phones and service may slow its growth for awhile. At the same time, the shift of some cellular subscribers from analog to GSM may actually free up some analog capacity for new cellular users. (Motorola is the major base station equipment supplier for Cellnet; Orbitel/Ericsson supplies Vodafone.)

The total subscriber base in the U.K. was expected to reach almost 1.6 million by the end of 1993; about 17,800 of them GSM, the rest analog, for a total penetration of 2.8 percent. These numbers are expected to double by 1997 to 3 million cellular subscribers.

Together, the United Kingdom, Germany, and France are expected to account for close to half of the GSM subscriber market by 2002. Germany already represents the fastest-growing area for GSM with an estimated 100,000 GSM phones in use at the end of 1992. Although many of those subscribers were part of GSM field trials, Nokia, Europe's largest cellular equipment manufacturer, expects Germany to account for fully one-third of the GSM subscribers in Europe by the end of 1993. Some analysts expect the German cellular market to reach 640,000 analog users and slightly more than 2.9 million GSM subscribers by the end of 1997. That's a total penetration rate of about 4.5 percent.

One of the striking features of the European and Asian cellular markets is the level of foreign participation. It's a mixed bag with heavy emphasis on international alliances. Examples of licensees by country, system, partners, and their percentages of ownership, follow:

DENMARK

Dansk Mobiltelefon

BellSouth (U.S.), 29 percent

GN Greater Northern (Denmark), 51 percent

NordicTel (Sweden/U.K.), 20 percent

FRANCE

Société Française de Radiotéléphone (SFR)

Compagnie Générale des Eaux (France), 42 percent

BellSouth (U.S.), 4 percent

Vodafone (U.K.), 4 percent

Fabricom (Belgium), 25 percent

Magneti Mareli (Italy), 25 percent

GERMANY

E-Plus

Thyssen AG (Germany), 28 percent

Veba AG (Germany), 28 percent

BellSouth (U.S.), 21 percent
Vodafone (U.K.), 16 percent
Caisse des Dépôts (France), 2 percent
Other German companies, 5 percent

Mannesmann Mobilfunk

Mannesmann AG (Germany), 51 percent
Pacific Telesis (U.S.), 26 percent
Deutsche Genossen Bank (Germany), 10 percent
Lyonnaise des Eaux (France), 8 percent
Cable & Wireless (U.K.), 5 percent

GREECE

STET (the holding company for Italy's PTT), 100 percent

Panafon

Vodafone (U.K.), 45 percent
France Télécom (France), 35 percent
Intracom (Greece), 10 percent
Data Bank (Greece), 10 percent

JAPAN

TU-KA Cellular Tokyo

Nissan (Japan), 26 percent
DDI (Japan), 26 percent
Motorola (U.S.), 8 percent
GTE (U.S.), 3 percent
US WEST (U.S.), 2 percent
NYNEX (U.S.), 1 percent
BT (U.K.), 5 percent
Sony (Japan), 5.5 percent
Hitachi (Japan), 5.5 percent

Rogers Cantel (Canada), 2 percent

Various Japanese companies, 16 percent

TU-KA Cellular Kansai

Nissan Motors (Japan), 34 percent

Kobe Steel (Japan), 9 percent

Hitachi (Japan), 9 percent

BT (U.K.), 5 percent

Marubeni (Japan), 5 percent

Matsushita (Japan), 6.5 percent

NYNEX (U.S.), 2.0 percent

GTE (U.S.), 1.5 percent

Motorola (U.S.), 0.25 percent

Other Japanese firms, 27.25 percent

Kansai Digital Phone Co.

Japan Telecom Ltd. (Japan), 27 percent

Pacific Telesis (U.S.), 13 percent

West Japan Railway (Japan), 12 percent

Toyota (Japan), 11 percent

Cable & Wireless (U.K.), 7.2 percent

Various Japanese companies, 29.8 percent

Tokyo Digital Phone Co.

Japan Telecom Ltd. (Japan), 29.5 percent

Pacific Telesis (U.S.), 15 percent

Tokai Railway (Japan), 12 percent

Metrophone Group (Japan), 12 percent

Cable & Wireless (U.K.), 8 percent

Toyota (Japan), 4 percent

Nippon Steel (Japan), 4 percent

Other Japanese firms, 15.5 percent

NORWAY

Netcom GSM

Comvik (Sweden), 33 percent

Orkla Borregaard (Norway), 67 percent

PORTUGAL

Telecel

Espirito Santo (Portugal), 31.25 percent

Amoria (Portugal), 31.25 percent

Pacific Telesis (U.S.), 23 percent

Efacec (Portugal), 6.25 percent

Centrel (Portugal), 6.25 percent

LCC (U.S.), 2 percent

SWEDEN

NordicTel

(Sweden), 100 percent

Originally, the German government took a monopolistic approach to cellular communications, licensing the Deutsche Bundespost Telekom (DBT), Germany's national public phone network, as the only analog cellular service operator. More recently, however, the German government launched a major modernization program called Telekom 2000 to upgrade the communications network in eastern Germany and link it with the rest of the country. To do that, Germany has issued two GSM licenses. One went to the DBT, the other went to Mannesmann Mobilfunk GmbH , a privately held consortium, whose members include several European and U.S. telecom organizations (PacTel International has a 26 percent interest). The difference is that Mobilfunk's license requires that the network be based exclusively on the GSM standard. DBT's license covers both analog and digital cellular service.

Mobilfunk began its D2 Privat service in June 1992; the DBT system, called D1 Privat, began operation in July 1992, and both services have grown steadily. Mobilfunk predicted that it would have 350,000 subscribers by the end of 1993, a figure it surpassed in September. Typical of fast-growing cellular services, prices of vehicle-installed GSM phones dropped from about $1,563 in June 1992 to $812 in February 1993. Transportable models cost slightly more, and the first hand-held portables, which arrived in late 1992, were priced at about $1,406.

In February 1993, Germany issued a third license to the German consortium E-PLUS for a new national service based on the DCS 1800 standard. E-PLUS plans to begin rolling out its service in Berlin and Leipzig, the largest cities in eastern Germany, in 1994. Consortium members are led by Thyssen AG and Veba AG, each with a 28 percent stake in the organization. Other investors include BellSouth Enterprises of the U.S.; Vodafone Group Plc of the U.K.; and a number of small- and medium-sized companies from eastern Germany and the Caisse des Dépôt et Consignations Group of France, which operates the Cofira telecom network in France with BellSouth, and Bau GmbH, Industriemontagen Leipzig GmbH, Minol Mineralolhandel AG, and Part'Com S.A.

On the equipment side, E-PLUS has awarded Nokia of Finland the initial contract to supply the network infrastructure for the DCS 1800 PCN. The contract is valued at $100 million and covers only the first phase of the nationwide network. By the end of 1995, approximately 88 percent of the population of eastern Germany will have access to the network.

Thyssen officials have projected a three-year start-up loss, but believe they can sign up 10 million users by the year 2000 with annual fee revenues at that point of $11.6 billion. Germany's telepoint service, called Birdie, is another story. Like the U.K. and other regions, telepoint has generated little consumer interest, and the German government has stated that no additional licenses will be issued for cellular or cellularlike services through 1996.

France's France Télécom, which already operates the Radiocomm 2000 analog cellular system throughout the country, began operating its GSM network, called Itineris, in 1992 in Paris and Lyon. The country's second analog carrier is SFR, a private firm. GSM sales are expected to boom in 1994, and by 1995 the entire country should have access to GSM service. Initially, GSM phone service in France cost $70 per month, with a flat fee for calls ranging between 50 and 90 cents, depending on the location. EMCI, the market research organization, estimates there will be 900,000 GSM subscribers in France by 1996.

Alcatel, Matra, and Ericsson are the lead equipment suppliers for France Télécom. SFR is using equipment supplied by Alcatel, Siemens, and TRT, the French Philips subsidiary. Sales strategies for network equipment differ in France from those of end-user equipment. Markets for network equipment are controlled to a much greater degree by the PTT and by traditional PTT suppliers. In this market segment, some form of strategic alliance with France Télécom or a key supplier is almost required. However, American companies have done fairly well in France; BellSouth has acquired a 12.5 percent interest in France Télécom Mobiles Data, a new unit of France Télécom that plans to build a mobile data network in France.

Together, BellSouth and France Télécom plan to invest $80 million in the network, which was scheduled to begin service in Paris in 1993. BellSouth expects every city in France with more than 100,000 residents to have access to the service by 1997. Another American company, US WEST, owns 9 percent of La Lyonnaise Communication. Other U.S. companies that do business with France Télécom include Motorola, AT&T, Timeplex, Raychem, MCI, Hewlett-Packard, and IBM.

France's version of the cordless telephone, called Bi-Bop, is another telepoint service based on CT-2 technology. Launched in April 1993, the phone allows calls to any number provided the caller is no more than about 200 meters distance from a base station. At last count, Bi-Bop only had about 2,000 subscribers, and most of them were participants in the initial trial of the system.

Ameritech International, a subsidiary of Chicago-based Ameritech, will help build and operate a GSM-based cellular network in Norway, one of Europe's leading per capita users of mobile phones because so many Norwegians own second homes, few of which have fixed telephone service. Ameritech International and Singapore Telecom have won government approval to acquire 49.9 percent interest in Netcom GSM, a Norwegian firm. The three companies are working together to build and operate the cellular network in Norway. The system was scheduled to be operational by the end of 1993.

JAPAN'S CELLULAR USE TO MUSHROOM

	Subscribers (thousands)		
	ANALOG	DIGITAL	TOTAL
1990	780	0	780
1991	1,170	0	1,170
1992	1,640	0	1,640
1993	2,200	100	2,300
1994	2,260	600	2,860
1995	2,300	1,280	3,580
1996	2,200	2,100	4,300
1997	2,000	3,160	5,160
1998	1,900	4,300	6,200
1999	1,800	5,630	7,430
2000	1,600	5,960	7,560

Source: Ministry of Posts and Telecommunications

Japan's cellular market is expected to grow rapidly now that several new network operators are competing with Nippon Telegraph & Telephone (NTT).

Switzerland appeared to have made an early commitment to GSM, but may change that decision. The Swiss PTT has been testing QUALCOMM's CDMA digital cellular system against GSM and its existing analog system. When last heard from, Swiss officials would say only that they were impressed with CDMA.

Clearly, the Far East is going to be a very large market for all segments of wireless personal communications, but some regions are already ahead of others in the implementation of GSM and other services. EMCI expects the Asia-Pacific region to grow from 4.1 million cellular subscribers in 1992 to 16.9 million in 1997. With the introduction of a private carrier and continued high economic growth, EMCI also looks for rapid subscriber growth in South Korea, adding 1.5 million subscribers by 1997.

Other areas, such as Australia, Thailand, Taiwan, and Hong Kong, will also benefit from new digital cellular systems. By 1997, EMCI projects that 41 percent of the Asia-Pacific subscriber base will be using digital systems. GSM and a digital Advanced Mobile Phone System (AMPS) are expected to make up the remaining digital subscriber base. However, even with the onslaught of new digital systems, EMCI expects analog AMPS and Total Access Communication Service (TACS), one of several analog cellular systems in wide use in Europe, to account for 45 percent of the base in 1997.

Analog cellular was introduced in Japan in 1979, almost four years ahead of the United States. Nippon Telegraph & Telephone Co. (NTT), Japan's government-owned public corporation, had the cellular market to itself until 1985 when Japan enacted the Telecommunications Business Law, which essentially abolished the legal monopolies held by NTT, the Telegraph Public Corp., and Kokusai Denshin Denwa (KDD) and privatized the NTT Public Corp. In 1986, the Ministry of Posts and Telecommunications (MPT), which regulates the cellular industry in Japan, licensed two new service providers, Nippon Ido Tsushin and Daini Denden, Inc., to compete in the cellular market with NTT. However, neither company received a national license similar to NTT's. Ido—whose backers include Toyota, NEC, the Japan Highway Authority, and Tokyo Electric Power—is licensed to operate only in the Tokyo-Nagoya area. DDI, which is made up of eight affiliated Japanese cellular companies, can operate only in the remaining, mostly residential suburban, regions of the country. (One big difference between the U.S. and Japanese cellular markets is that in Japan, cellular phones are leased. However, that will change in April 1994 when the MPT will allow subscribers to buy cellular handsets at retail.)

Japan's digital mobile communication system is called Japan Digital Cellular (JDC). It is a TDMA-based system in the 800-MHz and 1.5-GHz bands, and therefore similar to the American TDMA network, but with one major exception; it will not be dual-mode (analog and digital).

Customers will be able to choose among four providers of digital cellular service. NTT and the new entrants will be licensed to offer digital cellular service nationally, although they are expected to begin operations on a regional basis. Ido and DDI also will be licensed to offer digital service, but only in their current regions.

No foreign manufacturer is producing or assembling cellular phones in Japan. The imports of cellular phones into Japan are limited to Motorola's sales to NTT and DDI. According to the *Japan Market Share Encyclopedia*, Mitsubishi led sales of car and portable cellular phones to NTT in 1991 with 200,000 units. Matsushita was second (150,000 units), followed in order by NEC (120,000), Fujitsu (80,000), Japan Wireless (27,000), and Motorola (8,000). EGIS, a research and consulting group that follows Japan's telecommunications markets, believes that new entrants in the Japanese market will wait for the implementation of digital cellular systems and the liberalization of the terminal market before attempting any new import activity.

NTT has already selected Motorola, AT&T, and Ericsson along with six Japanese manufacturers to develop its digital cellular network, and Fujitsu, Matsushita, Mitsubishi Electric, and Motorola were named to supply NTT with digital phones. Motorola, NEC, and Ericsson will also supply DDI. Ido has selected AT&T, NEC, Fujitsu, and Nokia Mobile Phones as its key equipment suppliers. Also, Ericsson and Toshiba of Japan have formed a joint venture to develop Ericsson's digital cellular equipment business in Japan. Under their agreement, Ericsson will supply Toshiba with equipment valued at $250 million or more for new digital cellular networks in Tokyo, Osaka, Kobe, and other Japanese cities. The networks are to begin operations by mid-1994.

Another emerging service in Japan is the two-way (send and receive) PCS-type Personal Handy Phone (PHP), or Personal Handy-Communications System (PHS), which will look very much like most cordless phones. Operating in the 1.9 GHz band, PHP field trials began in the fall of 1993, and commercial service is tentatively scheduled to begin in 1995. PHP reportedly will be offered through private networks and will not be subject to the same foreign ownership restrictions applied to common carriers. As a result, the PHP market may be open to more competition than cellular.

DDI says its new PHP will cost only $16.26 a month in access charges and 24 cents for 3 minutes of airtime. DDI also says its PHP base stations may only cost $30,000. But with a range of only 100 to 200 meters, more base stations will be required than for the typical cellular system. Ultimately, DDI envisions PHPs operating building to building with base stations in homes, office buildings, and stores.

Japan has high hopes for PHP. The MPT is predicting that the low cost of PHP could push sales to 40 million units by 2010. DDI is also trying to introduce PHP to Korea, Hong Kong, Singapore,

Taiwan, and Thailand. DDI officials have also met with several American companies, including the regional Bell companies, cable television companies, and MCI Communications, in an effort to promote PHP as an international standard.

Hong Kong is the largest per capita market for cellular phones in the world, and the most competitive, and CT-2 is far more successful in Hong Kong than in any other area. The government introduced cellular in 1983 by granting Hutchison Telephone Limited, Pacific Link Communications Limited, and Hong Kong Telecom CSL Limited licenses to operate cellular networks. But three carriers were not enough to satisfy Hong Kong's "telephone fever." Capacity limitations created a slump in cellular handset sales in 1992, forcing the Hong Kong Telecommunications Authority to license a fourth cellular system. The telecom authority also ordered the three existing cellular operators to switch their systems from analog to digital by the middle of 1995. To protect their existing customer base and allow continued roaming into the PRC, they most likely will switch to dual (analog/digital) systems. Building the fourth digital network and switching the other three to digital will cost an estimated $150 million for telecom-

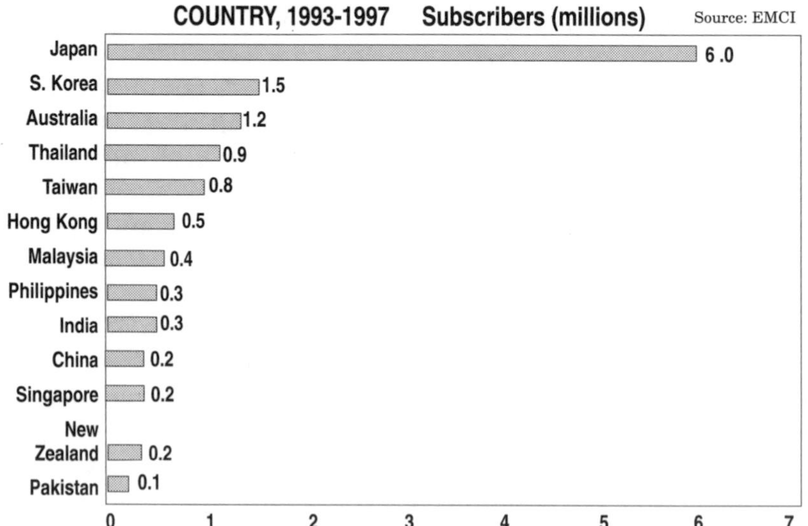

The Asia-Pacific region is projected to be the most significant source of cellular subscriber growth in the mid-1990s, growing from 4.1 million subscribers in 1992 to 16.9 million in 1997.

munications equipment, not including handset sales. The Hong Kong telecom authority anticipates the need for a fifth cellular network operator in 1995.

The U.S. and Japanese cellular manufacturers dominate the Hong Kong market, with the U.S. share estimated at 40 percent. Motorola is believed to have the largest market share because of an exclusive supplier agreement with Hutchison Telephone Limited. Most of the remaining equipment market is controlled by several Japanese producers and some European manufacturers, led by Ericsson. Andrew Corp. got a piece of the market through a $2 million contract with Hongkong Telecom CSL to build an underground cellular network for the Hong Kong Mass Transit Railway. Hong Kong Telecom CSL, the local and international telephone carrier in Hong Kong, operates the largest single cellular network in Hong Kong, with extensive roaming arrangements throughout China, and is expected to introduce a new GSM network in Hong Kong.

Unlike the United Kingdom, CT-2 started out very strong in Hong Kong. British CT-2 operators were never able to establish an adequate number of telepoints throughout the large land area of Britain, greatly limiting the range and success of their services. In Hong Kong, however, CT-2 network operators can serve the entire territory, an area covering less than 1,000 square kilometers, with far fewer telepoints. Also, most Hong Kong residents walk or use public transportation instead of personal vehicles and would more likely use a CT-2 phone. Expectations for CT-2 were high enough in Hong Kong that the Hong Kong telecom authority has issued four CT-2 licenses—to Hutchison Paging Limited, Chevalier Limited, Personal Communications Limited, and Hong Kong Callpoint Limited. Each of these operators was expected to spend $50 million to establish their networks.

The CT-2 operators and manufacturers had forecasted a boom in the Hong Kong market within the next decade. Market sources based their optimism on the stunning growth of the local cellular and paging markets. CT-2 sales were expected to exceed the 40 percent annual growth of the cellular telephone and eventually serve more than the 800,000 pager subscribers. Hutchison and Chevalier projected the CT-2 market to grow to 500,000 by the end of 1995, with CT-2 telepoints covering the entire territory.

Hong Kong's cellular operators never really agreed with those projections. They believed the success of CT-2 is due to the saturation of the local cellular market. Their view was that the introduction of a fourth cellular network and the shift from analog to digital cellular networks will lure most prospective portable telephone users back to cellular. The CT-2 camp claimed it was targeting a different sector of the market—the price conscious mainstream consumer. With an average price of $300 per phone and an average monthly charge of $23, the overall cost of a CT-2 unit is about one-fourth the price of a cellular phone.

To date, most of the early CT-2 handset, homebase, and telepoint units have come from Motorola, GPT Limited, and Shaye Communications Limited. At the end of 1992, Motorola, through an exclusive agreement with Hutchison, controlled at least 90 percent of the market for CT-2 handsets, homebases, telepoints, and other CT-2 operations-related equipment in Hong Kong. GPT and Shaye shared the remaining 10 percent. In time, other telecom equipment manufacturers, such as Sony, Samsung, Kenwood, and Panasonic indicated they might enter the CT-2 equipment market. Based on recent market developments, they may change their positions; at one point, Hutchison had forecast that CT-2 would contribute nearly HK$200 million in profits by 1995. In mid-1993, the company said it would more likely break even. Then, in November 1993, Hutchison Telecommunications announced that it would close its Rabbit telepoint service (in the U.K.) and sell its mobile data operation. However, Hutchison said that it would continue to develop its personal communications network.

The People's Republic of China (PRC) is another potentially huge market opportunity for wireless. At least 1 million households in China are on waiting lists for telephones even though the cost of installing a phone in the PRC is $925. Because phones are so hard to get, visiting businesspeople or foreign companies in residence usually rent cellular equipment. Motorola has a major position in the PRC, but Sweden's L. M. Ericsson is the current market leader for cellular subscriber units and has won a $150 million contract from China for cellular base stations. Another supplier, Novatel-Fonic Asia Communications Corp., a joint venture of Canada's Novatel Communications Ltd. and Fonic Inc., a Hong Kong–based consumer electronics company, has won orders to construct two cellular phone systems in China. Novatel-Fonic is also working with China's Ministry of Machine-Building and Electronics Industries to produce cellular phones, pagers, and CT-2 phones. The venture will set up two analog cellular systems in the north and central regions of China. Chengdu Tongfa Telecommunications, a cellular operating company organized by Sichuan PTT, has selected Hughes Network Systems to build a digital portable cellular service in Chengdu, the capital of China's Sichuan Province. The initial contract, valued at $15 million, will serve more than 20,000 subscribers, but that will be expanded eventually to handle 100,000 subscribers.

Economic growth in China has forced the country's telecom authorities to double their expectations for telecommunications services. Plans now call for the installation of 31 million telephones by 1995 and 65 million by 2000. Most of this development will be focused on major cities and is expected to generate big growth numbers for cellular and, eventually, PCN equipment and service suppliers.

In Taiwan, the government-run Computer and Communications Laboratory in Hsinchu is coordinating a $2.8 million effort to develop DECT-based hand-held and PBX products. Estimates run as high as 42 million DECT users in the country by the year 2000. Six companies are taking part in writing technical specifications for the new products: Tatung Co., Taiwan Telecommunication Industry Co., Sinoca Enterprise Ltd., Taicom System Ltd., Global Communications Inc., and Myday Technology Ltd. All are based in Taipei.

In South Korea, which introduced mobile communications in 1984, the state-run mobile telecom carrier, Korea Mobile Telecommunications Corp., had 340,000 subscribers in 1992. That figure is expected to grow to 5 million by the year 2000, despite the fact that Taehan Telecom Ltd., a consortium led by the Sunkyong Business Group, has backed out of its winning bid for Korea's second cellular phone carrier license. In addition to Sunkyong, the consortium includes GTE, Vodafone, and Hutchison Whampoa Ltd. (Sunkyong was charged with nepotism—the son of Sunkyong's chairman is married to Korean President Roh Tae Woo's daughter. Despite strong protests from GTE and Hutchison, South Korea's political leaders decided to let the next government select the nation's first private mobile phone operator.)

The Korean Ministry of Communications has already selected QUALCOMM's proprietary CDMA digital technology for the Korean cellular telephone system. The ministry has set a 1995 date for commercial CDMA service, advancing the schedule for digital cellular service in Korea by two years. The schedule now calls for prototype equipment in September 1994, followed by commercial field trials in early 1995 and commercial service later that year.

The equipment for the Korean network will be supplied by four Electronics and Telecommunications Research Institute (ETRI)–designated Korean manufacturers: Goldstar Information & Communications Ltd., Hyundai Electronics Industries Co., Maxon Electronics Co., and Samsung Electronics Co. Under license from QUALCOMM, the developer of CDMA technology, Maxon will produce subscriber equipment only, while the other three companies will develop both subscriber and infrastructure equipment. In addition to providing equipment for the Korean market, these manufacturers will become alternate sources of CDMA equipment for networks in the United States and other countries implementing CDMA.

Eastern Europe and the former Soviet republics, once restricted from buying Western high-technology products, have become major customers of virtually every major telecommunications equipment manufacturer. Rather than spend years trying to bring their telecommunications systems up to the standards of the rest of the world by installing new telephone lines, most of these economically depressed areas have turned to wireless systems. In the reunited eastern and western states of Germany, for example, the Bundespost is implementing wireless local-loop services to provide immediate, short-term telecommunications services. One of the key features of this system is that international calls are possible without having to go through the fixed network.

In general terms, this change in trade policy means that the United States and its allies can make better quality equipment available for export to eastern Europe and the newly independent states than was possible in the past. It also creates a major new market opportunity. According to the North American Telecommunications Association's (NATA) 1992 *Telecommunications Market Review and Forecast*, "There is every reason to look covetously at the Eastern European telecommunications market. With antiquated telephone networks, Eastern European countries, including the former Soviet Union, may spend up to $150 billion over the next decade to upgrade their networks. Taken together, these countries may eventually represent 20 percent of the world market for telecommunications equipment."

Not all eastern European and new Commonwealth of Independent States (CIS) countries are equally developed in terms of phone service. The CIS has the third largest telephone system in the world, with 27 million access lines, but has a very low teledensity of 9 phones per 100 inhabitants. What was Czechoslovakia probably has the best telephone system in the region, despite a relatively low penetration of 25 phones per 100 people. Hungary's phone system is not nearly as good, but it is making rapid progress. The NATA report suggests that companies with an aggressive, successful strategy stand a good chance of breaking into the eastern European markets as trade between the East and West grows and becomes less complicated politically.

QUALCOMM fits the description. Despite what appeared to be certain commitments by Russia to adopt GSM as a digital cellular standard, QUALCOMM has signed a memorandum of understanding with the Telecommunications Ministry of the Russian Federation and with Russ Telecommunications Co., a U.S.-R.F. shareholding company, to immediately begin establishing CDMA as a new wireless standard in Russia. The agreement includes a plan aimed at the allocation of spectrum for a CDMA network. The system would be based on the IS-54 standard published by the U.S. Telecommunications Industry Association. The Russian ministry said it was ready to promote the deployment of a CDMA-based wireless local-loop network to serve at least 20,000 users by the end of 1994.

Frost & Sullivan/Market Intelligence Research Corp. believes the near-term opportunities for wireless communications may be even better in eastern Europe than in the U.S. because the area has fewer analog lines, making digital networks easier to establish. MIRC studies also indicate that when eastern Europe develops its own modern telecommunications infrastructure, manufacturing will shift slowly from foreign facilities to either native companies or foreign companies with local subsidiaries. If that occurs, labor costs and customs duties will eventually be cheaper within eastern Europe. In fact, several companies are now building or refurbishing factories within eastern Europe in anticipation of those changes.

Hughes Network Systems, along with GTE Spacenet International, SEL Alcatel, and San Francisco/Moscow Teleport, Inc.

(SFMT), a little-known, New York City–based consulting firm, are installing what could become the first large-scale residential wireless telephone system in the world in Tatarstan, a Russian republic 500 miles east of Moscow with 3.6 million residents. Hughes' initial contract with Tatarstan is worth $48 million and calls for the company to provide hand-held and mobile phones, using its E-TDMA digital cellular telephone technology. SFMT is the general contractor for the project, GTE Spacenet International is building the earth station, and SEL Alcatel is providing the digital base station switches. The system is expected to be completed in early 1994, but that schedule could easily slip as Tatarstan and Russian officials have been arguing over who controls spectrum allocations in the area, and who should operate the system once it is completed.

US WEST NewVector is part of a group building a cellular system in Moscow. The company's investment is relatively small ($18 million over two years), but it helped establish US WEST's position in the region. US WEST also operates a cellular network in St. Petersburg and owns 49 percent of a cellular project in Budapest, Hungary, with equipment supplied by L. M. Ericsson. Another U.S. cellular carrier, Bell Atlantic Mobile, operates a cellular network in the new Czech republic to supplement that country's wired phone service. Another U.S. firm, Plexsys International, has formed a joint venture with Russian partners to install the first AMPS standard cellular phone system in Moscow. The local partner in the venture, known as Euronet, is Vimpel, one of Russia's (and the former USSR's) top weapon system designers.

Andrew Corp., a U.S. equipment supplier, is developing projects in Moscow, St. Petersburg, and Warsaw. In Moscow, Andrew is part of a joint stock company that is installing fiber-optic and two-way radio communications in the underground metro rail system. The radios will be used for emergency services. Andrew's Telecommunications Systems Group expects to collect revenues from users of the fiber network, which has already been tariffed by the local communications ministry. Initial customers will be mainly offices and hotels. Andrew has a similar arrangement in St. Petersburg. In Warsaw, Andrew has installed a two-way cable network, which will be tied into a new cellular phone system. Andrew is also involved in a joint manufacturing venture with

Iskra, a company in Krasmoyarsk, Siberia, a city with 1 million residents. Using Andrew machinery, the business alliance (called Anisco) started producing antennas in August 1992 for the Russian market; eventually, Andrew hopes to export the products.

Wireless pay phones also represent an excellent market opportunity in both eastern Europe and the CIS. Again, it's a matter of expediency. Most new pay phones in eastern Europe will very likely be cellular to get those countries up to speed with the rest of the world as quickly as possible. The new pay phones will look very much like the wireless, solar-powered emergency call boxes along the side of many U.S. highways. Initially, the pay phones may be analog because the telecommunications authorities in these countries don't want to wait until digital pay phones become readily available.

Getting paid for some of these new installations could be a problem, however. At the end of 1992, London-based Frost & Sullivan International estimated that the former Eastern Bloc countries faced a $50 billion shortfall in funding their telecommunications projects. At the time, FSI indicated a total of $80 billion would be needed up to the year 2000 to pay for telecommunications hardware in Poland, Czechoslovakia, Hungary, Romania, Yugoslavia, and Bulgaria.

Cellular service is also now available in every region in Mexico. According to the CTIA, nationwide roaming became available in 1992, and Mexico has signed international roaming agreements with the United States. By the end of 1992, the CTIA says the cellular industry in Mexico was serving between 250,000 and 300,000 subscribers. With a population estimated at close to 86 million, Mexico has plenty of room for cellular growth.

In Canada, cellular systems operate on the same frequencies as the U.S., but they are licensed differently. According to the CTIA, Block A, the 800-MHz service, is provided by a single, nationwide carrier—Mobility Canada—while Block B, the 900-MHz service, is provided by provincial carriers. Canada's 1992 population was 27.2 million and the CTIA estimates that roughly 85 percent have access to cellular service. According to Canada's Department of Communications, approximately 1.02 million people subscribe to cellular service.

Also, in December 1992, Canada awarded four consortia licenses for PCS while Mobility Canada won a national license to provide a personal cordless telephone (PCT) service, using CT-2 Plus. (An enhanced version of CT-2 developed by Northern Telecom Ltd. of Canada, CT-2 Plus features an optional dedicated common signaling channel, cellularlike handoff, enhanced transmission speeds of 19.2 kilobits per second, subchanneling to allow capacity increases as the technology improves, improved security, and integrated voice, data, and imaging.) Mobility Canada's PCT public zone service will be launched in major Canadian city centers in 1994 in stadiums, hotels, shopping malls, airports, factories, offices, and public buildings. PCT private zones would cover the area in and around a customer's home or office. The service should be available to approximately 18 million Canadians by 1998.

Owners and/or shareholders of Mobility Canada include AGT Cellular Limited, BCE Mobile Communications Inc., BC Tel, Edmonton Telephone Corp., Island Telephone Co., Manitoba Telephone System, Maritime Telegraph & Telephone Co., New Brunswick Telephone Co., Newfoundland Telephone Co., Quebec Telephone Co., Saskatchewan Telecommunications, and Thunder Bay Telephone.

Mobility Canada believes the market for PCT network and user equipment in Canada will exceed $1.5 billion between 1993 and 1998.

The four newly licensed PCS companies are Canada Popfone Corp., Mobility Personacom Canada Ltd., Rogers Cantel Mobile Inc., and Telezone Inc. They will use the 944–948.5-MHz band and expect to begin service in 1994.

Other areas of the world are fairing just as well. In Brazil, for example, Telesp, the state-owned telecommunications service provider in São Paulo, announced at the end of 1992 that NEC of Japan had won three tenders to supply equipment for a new mobile cellular network in the state. The contract is expected to give NEC 70 percent of the cellular equipment market in Brazil. Telesp has been evaluating bids in separate tenders to supply equipment for other regions of Brazil, including some of the interior states.

CHAPTER 7

PAGING/WIRELESS MESSAGING

Here Come the Road Warriors

One of the more revealing illustrations of just how far we have come with wireless technology appeared in a *New York Times* article on the new AT&T/EO Personal Communicator, which allows you to write with a special pen on an electronic screen and then transmit your handwritten message to another pen-based terminal. The message displayed on the screen of the device reads, "Mr. Watson, come here, I want you." Still, at $1,600 (the cellular phone attachment is extra), the EO 440 is an expensive way to send a message, even for the most ardent wirelessphile.

Actually, pagers, or "beepers," continue to be the most popular form of mobile communications available today. They're small, reliable, and relatively cheap. About 15.3 million of them are in use in the United States. Almost 40 million people are walking around somewhere in the world with a one-way pager, and the pager industry is projecting more than 50 million subscribers by the year 2000.

But paging will soon represent only a small part of the total wireless messaging market. The introduction of communications-capable laptop, palmtop, notebook, and subnotebook computers, and what *Esquire* magazine calls "digital doodlers"—portable personal communicators, personal digital assistants (PDAs), personal intelligent devices (PIDs), and electronic organizers—will represent a huge advance in wireless messaging technology and will vastly expand the wireless market. Motorola believes 20–26 million of these wireless data devices will be in use by the year 2000. In terms of dollars, Motorola says it agrees with McLaughlin & Associates' projection of a $21.5 billion worldwide market in 2000.

How is this going to happen?

RadioMail, a pioneer in the field, sees wireless messaging growing from the convergence of three developments: (1) the boom in mobile computing created by relatively low-cost portable computers and personal communication devices, (2) the proliferation of mobile wireless data communications wide-area networks (WANs) and one-way paging networks, and (3) the availability of economical RF data modems that allow PCs to communicate remotely, or wirelessly.

A number of networks that support one-way wireless data communications are already in place, including Motorola's EMBARC, MobileComm's Nationwide Messaging, and Mobile Telecommunications Technologies Corp.'s SkyTel service. Two other systems, Advanced Radio Data Information Service (ARDIS) and RAM Mobile Data, offer two-way wireless communications. Both in the WAN segment of the market, RAM Mobile Data and ARDIS have emerged in a very short time as major forces in wireless communications.

Launched in 1989 without a single customer by RAM Broadcasting Corp., RAM Mobile Data can transmit short, two-way, public-switched, packet-switched data messages between portable computers.

RAM is based on Mobitex, an open, nonproprietary protocol developed by Swedish Telecom and Ericsson Radio Systems and maintained by the Mobitex Operators Association (MOA). Besides the United States, Mobitex systems are operational in Canada, the U.K., Finland, Sweden, and Norway. Licenses for two Mobitex networks have recently been awarded to France. The MOA is also working with network operators in other European countries, Latin America, and the Pacific Rim to launch Mobitex services in those areas.

Now 49 percent–owned by BellSouth Enterprises (BSE has invested more than $300 million in RAM), RAM Mobile Data operates in more than 200 U.S. markets. In addition, Mobitex networks have been installed in Canada, Finland, Sweden, and the

U.K. Other Mobitex networks are planned in Australia, Belgium, France, and the Netherlands. To facilitate wireless data communications between these areas, the MOA is developing international roaming capabilities for Mobitex users. This would allow a traveling executive (companies such as Southern California Edison and AGFA, for example, already subscribe to RAM's wireless E-mail service) to use RAM's services and portable computers equipped with radio modems to send and receive E-mail messages, from virtually anywhere, over both private and public systems, such as CompuServe, MCI Mail, and Internet. The MOA demonstrated international roaming in 1993 and plans to offer commercial service beginning in 1994.

The BellSouth and RAM partnership, announced in October 1991, includes RAM's mobile data operations in the U.K., as well as cellular, paging, and other holdings in the U. S. BellSouth has contributed complementary paging properties to the joint venture and provided more than $300 million equity funding to develop RAM's network in the U.S. and the U.K. and to pursue similar opportunities worldwide. RAM's other U.K. partners are France Télécom, Swedish Telecom, and Bouygues, a large French construction firm.

In operation, seamless roaming allows RAM subscribers wireless data access without special arrangements or codes—messages find mobile users automatically within RAM's coverage area. RAM transmits at 896–901 MHz and receives at 935–940 MHz. Data is sent over the RAM network in packets, very much like a cellular telephone system. But with no need for cellularlike handoff from cell to cell, there is virtually no risk of disconnected communications. Even if the user's equipment is turned off, messages are stored for retrieval on command. The capacity of the RAM network is virtually unlimited. RAM uses multichannel radio stations, which can simultaneously operate up to 167 channels. RAM has as many as 30 channels available for the larger metropolitan areas and can expand network capacity transparently to the subscriber by adding channels to existing base stations or by deploying additional base stations.

Field service dispatch is a typical application for RAM, but wireless credit authorization, remote meter reading, and energy management by utility companies are being added to the RAM wireless network. Minneapolis-based National Computer Systems was using paging to call its field engineers on service calls, but now subscribes to RAM, allowing the information management company to access centralized databases and contact other NCS employees directly. By replacing its private mobile data system with RAM, National Car Rental gets a national and less costly network with international service potential. Executives of Southern California Edison are now using RAM with palmtop computers and radio modems to stay in touch with their offices. And in a joint venture with Business Partners Solutions, Inc., a systems integrator, RAM has begun offering two-way wireless data communications for IBM AS/400 minicomputer systems users.

Several manufacturers have introduced or announced Mobitex-compatible portable and palmtop computers, radio modems, and other devices. Ericsson GE introduced a hand-held, battery-powered wireless modem. Motorola and AT&T have also announced Mobitex-compatible modems, and Mobitex radio modems will be available on a credit card–size PCMCIA module that fits into a slot in portable PCs. (The Personal Computer Memory Card International Association, or PCMCIA, was formed to develop and promote standards for memory cards and connecting slots for mobile computer products.) Motorola, RAM, and several software specialists are also developing application program interfaces (APIs) to make it easier for software packagers and systems integrators to develop new wireless data applications.

To quickly expand its services, RAM is now available through some 6,200 retail computer outlets, including Computerland Express, CompUSA, and Micro Center. Intel and Ericsson GE are jointly developing PC enhancement products for the RAM network, using Ericsson's portable radio technologies, and Intel and BellSouth are exploring new mobile computing products and services for the RAM network. RAM is also working with AT&T, Digital Equipment Corp., GO Corp., Lotus, Novell, and WordPerfect to integrate wireless messaging into their products and services.

ARDIS, the other two-way packet-switched data network, was inaugurated in early 1990 by combining the Motorola-developed private mobile data network used by IBM's National Service Division with Motorola's own shared-use radio data network. Today, ARDIS is a 50/50 joint venture of IBM and Motorola.

ARDIS currently holds FCC licenses for single-channel operation in most of the more than 400 MSAs nationwide. ARDIS transmitters are networked via dedicated land-based lines. The network transmits at 810–837 MHz and receives at 855–837 MHz.

Motorola designed, produced, and installed IBM's radio data network in 1983. When ARDIS was launched, the network was already being used by 16,000 IBM and 2,000 ROLM service personnel. So far, ARDIS has focused on vertical markets such as the field service industry. Guaranteed Overnight Delivery, which has G.O.D. emblazoned across the sides of its trucks, is a typical ARDIS user. The Kearny, New Jersey–based company offers overnight deliveries "or your money back." To help keep that promise, the ARDIS wireless data network was incorporated into G.O.D.'s delivery and pickup operations. Household Finance Corp., a consumer financial services firm, uses ARDIS and GRiD Systems' pen-based GRiDPAD for remote instant processing of consumer credit applications. Other users include Sears, UPS, and Avis Rent-A-Car.

Radio Frequency Data Network Systems, Inc., is another ARDIS customer, but also remarkets ARDIS' nationwide data radio airtime along with its specialized RF software. Customers receive airtime bills directly from ARDIS, and RF Data collects commissions as an indirect sales channel on monthly airtime revenues. RF Data has installed software products for ARDIS customers such as Otis Elevator and Pitney Bowes.

ARDIS has also moved into Canada through the creation of Bell-Ardis, a joint venture between BCE Mobile Inc. and Motorola Canada Ltd. Bell-Ardis has an agreement with IBM Canada to provide radio data service to its more than 1,000 field service personnel in Canada, mirroring a similar arrangement between ARDIS and IBM in the U.S. Bell-Ardis clients include ADP Automotive Claims Services Division's Audatex CD auto repair field estimating service and Otis Elevator, with more than 300 field technicians on the Bell-Ardis network.

To further enhance its place in the international market and help promote Motorola's packet data protocols as an international industry network standard, ARDIS has formed the Worldwide Wireless Data Network Operators Group. The group's original five members include ARDIS, Deutsche Bundespost Telecom, Hutchison Mobile Data (U.K.) Limited, Bell-Ardis, and Hutchison Mobile Data (Hong Kong) Limited.

Why not make ARDIS and RAM compatible? In fact, Motorola

and RAM are working on that; they're trying to develop an interface that would allow software written for ARDIS to work with the RAM network. Meanwhile, the competition is gaining on them, albeit slowly.

Mobile Telecommunications Technologies Corp. (Mtel) provides nationwide wireless messaging services through its principal subsidiary SkyTel Corp. In September 1991, SkyTel introduced SkyWord, the country's first nationwide alphanumeric paging service. A month later, the company announced SkyLink, the first "wireless mailbox," which uses AT&T's EasyLink Services. In early 1992, SkyTel introduced the Message Card, the only credit card–size paging unit available in the U.S. at the time. (A fast start, but SkyTel faces heavy competition from several new companies, including Arch Communications Group, Dial Paging Inc., Paging/ Network Inc., and Metrocall Inc. Arch is organizing a consortium of small, independent paging service providers across the country.)

In June 1993, the FCC granted Mtel a final "Pioneer's Preference" award for its so-called multicarrier modulation technology that can transmit 24 kilobits per second simulcast signal in a single 50-kHz channel, or about 10 times the bit rate of existing simulcast paging systems using equivalent bandwidth. Mtel also got the go-ahead to build a two-way nationwide wireless network (NWN). Once it receives a formal license from the FCC (virtually assured for companies with Pioneer's Preference status), Mtel says it will operate initially in the top 300 U.S. markets by 1995.

The NWN is a location-independent system, meaning that messages can be sent and received without actually knowing where they are going or where they came from. NWN will blanket U.S. metropolitan areas with a network of more than 3,000 base station receivers and transmitters, eliminating the need for users to go through inconvenient and costly roaming or location instructions.

The Mtel award was part of a larger but simultaneous FCC decision that allocated three bands for new, narrowband PCS systems—901–902 MHz, 930–931 MHz, and 940–941 MHz. These services, which have been known as advanced messaging services (AMS), may include two-way acknowledgment paging, advanced voice paging, electronic mail, and data communications. This was close to what the Personal Communications Industry Association (PCIA) had requested in its original petition, submitted to the FCC more than two years earlier. However, the PCIA had proposed that only a single band, 930–931 MHz, should be reallocated to AMS. By exceeding the PCIA's original request, the trade association said the FCC was obviously responding to the growth of the wireless industry and petitions submitted by carriers to build AMS systems.

The FCC decision provides for three types of narrowband PCS or AMS licenses—national, regional, and local. The regional licenses will be based on the 47 Major Trading Areas defined by Rand McNally, plus Alaska and Puerto Rico. The local licenses will be based on Rand McNally's 487 Basic Trading Areas.

As part of the same ruling, the commission ordered that cellular and local exchange carriers—the part of the national telephone network controlled by the local telephone operating company—may participate in narrowband PCS or AMS without any restrictions. The FCC's decision will allow NWN subscribers to send and receive wireless messages using portable computers and the newly emerging hand-held messaging units called personal digital assistants. Mtel is working with manufacturers to produce these devices, as well as PCMCIA card and chipset wireless modems to enable notebook computers, PDAs, and personal electronic organizer users to utilize the NWN.

Using an experimental license granted by the FCC in April 1992, Mtel has already constructed an NWN demonstration system in Dallas. This system enables users to send and receive messages while working in downtown buildings, traveling in moving vehicles, or simply walking down the street. The Dallas system was built using proprietary NWN messaging technology developed and implemented by Mtel and Motorola. Mtel hopes to work with telecommunications, software, and computer equipment companies to develop joint application and services for the NWN. The Silicon Valley–based venture capital firm Kleiner Perkins Caufield & Byers has a minority investment in the NWN, but industry observers believe it may take Mtel $500 million and five years to get the NWN up and operating, which is just about how long it took RAM to develop its system. Mtel says it expects the equipment for the NWN to be ready for commercial service in 1995.

SkyTel also is working with Hewlett-Packard Co. (HP) to jointly market the SkyStream, a 3-ounce wireless messaging receiver that enables owners of the Hewlett-Packard 95/100LX hand-held computer to receive text messages from SkyTel's satellite-based messaging network. Manufactured by Motorola as the NewsStream, the SkyStream operates on SkyTel's 931.9375-MHz frequency band for nationwide and international messaging services. It connects to the HP unit via the HP Mobile Data Link.

Wireless messaging has already become a boon for E-mail and fax service providers and users, radio modem manufacturers, and PCMCIA card developers. E-mail is the fastest-growing application for wireless messaging. According to the Electronic Messaging Association, there are 16 million E-mail users in the *Fortune* 2000 companies alone, and all these companies are primary candidates for wireless messaging applications.

A position paper developed by Ericsson GE Mobile Communications attributes much of the success of E-mail to the use of Internet, a series of networks that tie together public and private E-mail systems around the world. Through gateways between the RAM Mobile Data network and the Internet, provided by companies like RadioMail, users can access over 1 million host computers and 39,000 networks already connected to the Internet, including MCI Mail and a growing number of corporate mainframe and LAN-based systems. "The Internet," says the Ericsson GE report, "will do for mobile data what the wireline telephone network did for cellular telephone, allowing the first wireless customers to effortlessly communicate with the vast installed base of wired users." As more users go wireless, the E-mail market will grow precipitously.

Motorola's Electronic Mail Broadcast to a Roaming Computer (EMBARC) is a good example of the type of products and services that are being introduced for the wireless messaging market. EMBARC is a receive-only wireless electronic-mail service. Currently available in more than 220 markets in the U.S. and Canada, EMBARC uses Motorola's NewsStream receiver that connects to the serial port of most popular PCs, including Macintosh computers and HP 95/100LX palmtops. In addition to wireless E-mail, an application targeted to single users, EMBARC provides an easy and relatively low-cost means of updating an unlimited number of people with a single broadcast; for example, mass E-mail messages, updates on pricing schedules, weather and stock information, technical updates, inventory reports, pricing bulletins, and customer data can be broadcast to a work force scattered across the country. Using EMBARC, databases in portable computers can be updated automatically. EMBARC signals when messages are received, allowing the user to download the new information at any time. It can also receive and store up to 32Kilobytes of information when disconnected. Typical of the type of third-party add-on features available with EMBARC, a program called HeadsUp delivers news briefs to subscribers each morning from 24 information categories. Subscribers can get the full text faxed to them or dial into an EMBARC mailbox and download the information.

Motorola has made EMBARC accessible worldwide by signing interconnect agreements with several public E-mail providers—AT&T EasyLink, GE Information Services, IBM Mail Exchange, Pacific Bell, SprintMail, MCI Mail, and others. Tandy Corp. and Casio Inc., two major consumer electronics manufacturers, are developing PDAs featuring EMBARC's E-mail services, along with fax, calculator, and computer game functions. Called the Zoomer, the Tandy/Casio PDA is priced at $899.95. EMBARC hardware will add $395, plus a $15-per-month charge for the E-mail service. Incoming messages are free and unlimited; outgoing messages are priced based on their length and destination.

Motorola and MCI have also agreed to connect MCI Mail, which is now part of a larger, value-added service offered by MCI International, called MCI Global Messaging Services, with EMBARC. In the new system, messages can be sent to several people simultaneously from virtually anywhere in the world over MCI Mail to EMBARC recipients in the U.S. and Canada. EMBARC subscribers can then reply to MCI Mail users via traditional land-line access to EMBARC.

Lotus has introduced a series of one- and two-way wireless support products, including its E-mail product, the cc:Mail Wireless Pack, which enables mobile users to send and receive messages over the RAM Mobile Data network using an Intel Corp. wireless modem. cc:Mail offers LAN-based messaging into a single mailbox that can be accessed both wirelessly and over a phone line. Initially, Lotus is offering the cc:Mail Wireless Pack for use with both cc:Mail Remote for DOS and cc:Mail Mobile for the HP 100LX palmtop PC. Future versions of wireless cc:Mail will support both multisession configurations for IBM's OS/2 operating system and connection to RAM's networks via an X-25 leased line or public data network. These additional options will establish multiuser, simultaneous wireless connections for large groups of mobile cc:Mail users.

Lotus has also formed a marketing alliance with MCI to integrate cc:Mail into MCI global messaging services to market these combined capabilities worldwide. Under the agreement, cc:Mail Post Office will be incorporated into the MCI Mail network, enabling cc:Mail Mobile and cc:Mail Remote users to exchange messages without the need for a LAN or special gateway software on a personal computer. Users can also exchange messages with users of MCI and 54 other public E-mail services in 40 countries. The Lotus/MCI combo means that mobile data users such as field sales representatives can easily exchange messages with other cc:Mail users via MCI.

AT&T EasyLink is a global messaging service that offers electronic data interchange, gateways from LAN-based E-mail systems and telex, enhanced fax features, and information services. To further enhance their positions in wireless messaging, AT&T EasyLink Services and Intel Corp. have packaged a wireless modem from Intel with a new version of AT&T Mail Access PLUS software for Windows version 2.5. The modem recognizes the widely used Hayes AT communications command set, a feature that potentially opens up the use of the RAM networks to a wide base of AT-compatible software. The modems and software can be used to access the AT&T EasyLink Services through RAM's nationwide wireless network, which will provide the connection between AT&T Mail users and the EasyLink network. AT&T EasyLink users have access to 40 million E-mail users in more than 160 countries.

Motorola also is pushing the development of one- and two-way modems in the PCMCIA format and for good reason. Motorola plans to introduce a family of PCMCIA "intelligent" wireless modems covering three product categories: one-way wide-area networking, two-way wide-area networking (including cards for the ARDIS and RAM networks), and two-way local-area networking. All will be designed to support existing wireless communication protocols and networks.

In-Stat Inc. estimated that 121,000 of these cards were shipped in 1993, but that number will climb to more than 1 million in 1994 and jump to more than 9 million in 1997. Of the more than

100 types of portable products on the market, such as laptops and palmtops, close to half now offer a PCMCIA slot.

Even though the PCMCIA cards are about the size of a credit card and they all have 68 connector pins, there are actually three different types of cards. Type 1 cards are 3.3 millimeters thick and are used primarily for memory storage and software applications. Type 2 cards are 5.0 mm thick and usually hold modems or LANs for use in portable computers. Type 3 PCMCIA cards are 10.5 mm thick and are being used as hard disk drives. The next step is to ensure that all PCMCIA cards are interoperable—that they work the same in all mobile computing products, which is the goal of the PCMCIA.

How many PCMCIA cards can you carry? In time, that should cease to be a problem as more functions are embedded in the computers or communications devices themselves, or into new, more functional PCMCIA cards.

Where does all this leave plain old paging?

In fact, paging continues to grow an average 22 percent a year, mainly because of falling prices and a growing consumer market. Today, pagers are available in Sears, Wal-Mart, Radio Shack, Service Merchandise, Pep Boys, 7-Eleven, and many department and other broad-line retail outlets. At the same time, pagers keep getting better, and they are being offered with more features and in a variety of designs. Parents are giving their teenagers pagers so they can keep in constant touch with them. In Virginia, a garage operator hands out pagers to customers who would rather stroll a nearby shopping mall than sit and wait for their cars to be serviced; he pages them when the job is completed. The garage owner got the idea from a popular nearby restaurant that issues pagers to customers who prefer shopping at the same mall to sitting at the bar while waiting for a table. The maitre d' pages them when their table is ready.

Pagers can be purchased for less than $100, with a monthly service fee as low as $10. EMCI's studies indicate that the industry's revenue per pager continues to decline, but at a slower rate than in previous years. The industry's 1992 average revenue for a digital display rental pager was $15 per month, a slight decrease from $15.40 in 1991. The price reduction continues a trend—in 1989, display digital rental revenues per pager were $18.20 a month.

Pagers usually fall into two categories: the low-capacity type,

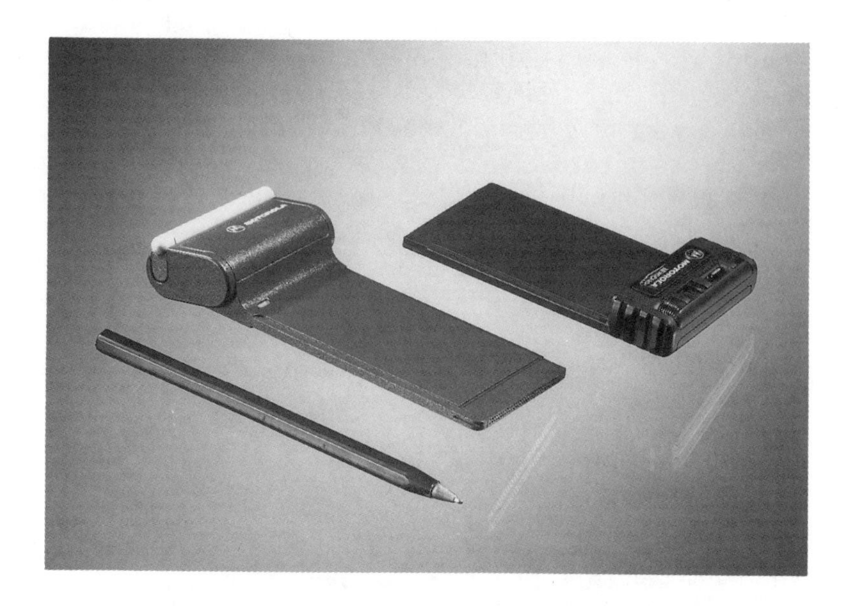

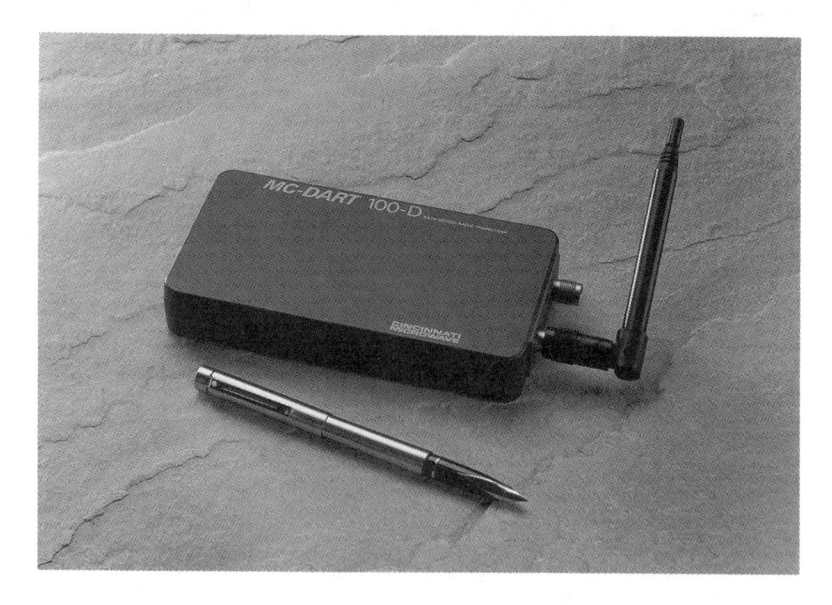

Despite their vastly different designs, the Motorola NewsCard, Cincinnati Microwave MC-DART 100-D, and Ericsson GE Mobidem (shown on p. 131) wireless modems all provide wireless data communications connectivity for industry-standard portable computers.

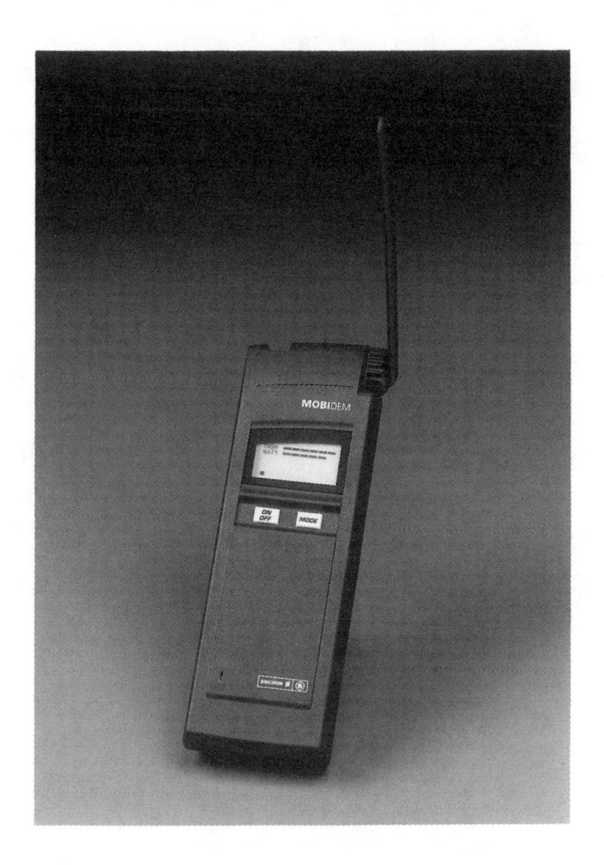

which are typically used in private systems such as hospitals and campus environments, and high-capacity radio systems operated by a local telephone company or radio common carrier (RCC). Low-capacity systems usually can support a subscriber base of no more than 200 users. In use, a pager is activated when a signal is sent to radio transmitters, which broadcast the signal over the coverage area. Each pager/receiver is tuned to a specific frequency and will activate only in its assigned frequency.

There are four major pager types: tone only (the familiar "beep" or vibration model), tone/voice (the beep plus an audible message), digital (an alert followed by a displayed telephone number to call), and alphanumeric (which can display a message of text and numbers). About 84 percent of the pagers in service in 1992 were digital, up from 80 percent in 1991. The PCIA says that only about 7 percent of the nation's pagers are alphanumeric, but

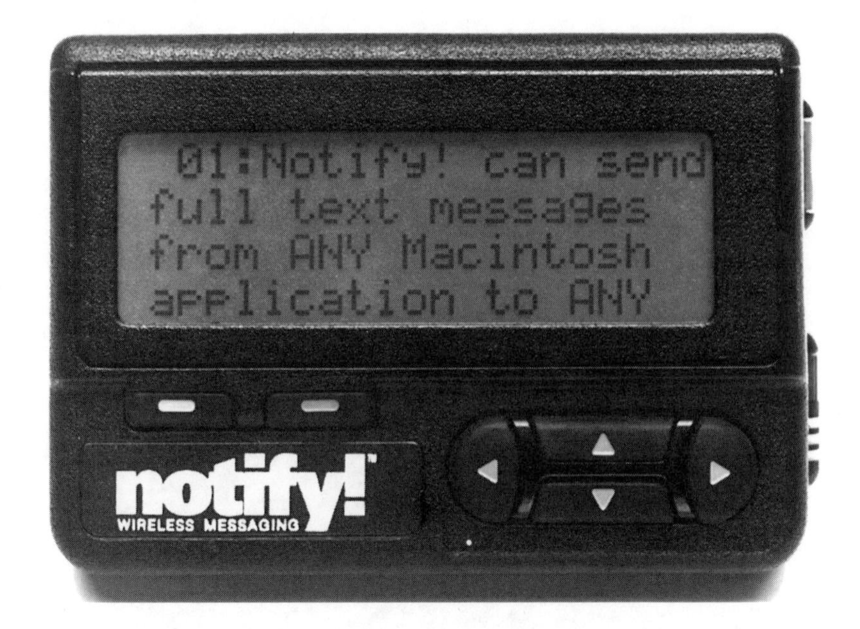

notify! is one of the more popular alphanumeric wireless messaging units on the market.

that sales of these units are growing at 27 percent a year, compared with 15 percent growth for digital models that display only a phone number to call.

Among pager manufacturers, Motorola is the dominant player, with an estimated 85 percent share of a $500 million market. NEC, which had a 14 percent share, left the market in April 1990, reportedly because of high product development costs. Panasonic is now believed to be a distant second to Motorola with about a 5 percent market share.

There is no shortage of new marketing ideas to expand the use of pagers. Metriplex Inc. offers pager users up-to-the-minute financial information and has introduced a service called Datapulse, which integrates Hewlett-Packard's HP95LX palmtop computer, a Motorola 3-ounce NewsStream wireless receiver, and Metriplex's software into a single package that permits traders in the U.S. to receive information 24 hours a day in several urban areas. Metriplex has also begun testing a service called LabAlert, which automatically displays patient test results for pager-equipped doctors. Then there is Las Vegas–based Beepers Plus Inc., which offers a service for gamblers with frequent updates of scores on sports events; Beepers Plus claims several thousand subscribers at $54 a month each.

The Minnesota Department of Transportation, the Federal Highway Administration, the University of Minnesota Center for Transportation Studies, and Motorola are working together to develop techniques for sending traffic and parking garage status information by pager. A typical message on an alphanumeric pager screen that is 20 characters wide and 4 lines deep might read

Delay—I-394 at Louis Av

Road construction

Left lane closed

or,

Garage A

Contract parking only

The Minnesota program hopes to make more detailed information available for notebook computer and PDA users.

MobileComm has introduced two national services that allow subscribers to send and receive messages, news reports, and electronic mail. Called CompuLink and MessageLink, the services can transmit messages to pagers, portable computers, and other personal communications devices. The services are available in more than 550 cities in the U.S., Canada, and the Caribbean.

Another paging service, Metrocall Inc., has introduced a nationwide paging network covering 78 major U.S. markets.

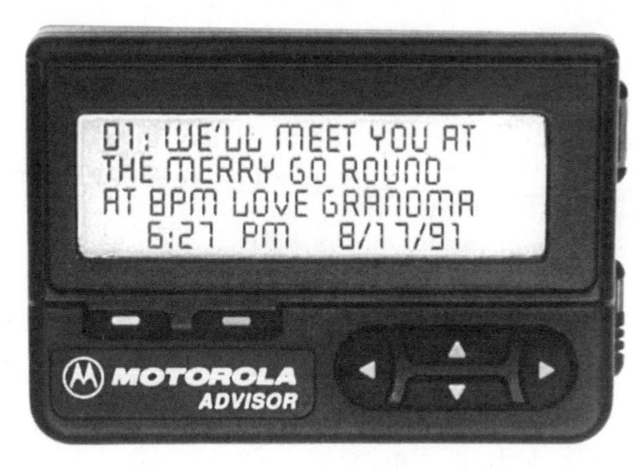

One of the more popular pagers available on the market today is the alphanumeric Motorola Advisor™.

To promote the use of pagers by nonbusiness and professional users, Motorola has developed a line of very colorful models, including green, neon pink, and even clear plastic. Motorola and Timex have been marketing a wristwatch/pager in selective markets since about the middle of 1990. MobileComm, a BellSouth company, markets the watch/pager, and Timex has also been working with American Paging to expand national retail distribution of its wristwatch/pager. (American Paging stocks pager telephone numbers and codes for each market where it operates or resells airtime for the Motorola/Timex unit.)

Motorola has also introduced FreeSpirit. Priced at $69 to $89, it emits a musical tone or slight vibration to signal the user and can store 10 messages in memory. Motorola's high-end unit is the Confidant; designed for executives willing to pay $179 to $199 for a pager that is the size of a credit card, it weighs 1.6 ounces; features a message overflow indicator, a clock, and a low-battery indicator; can hold eight messages; and emits a soft chirp. If you don't respond, it chirps again 5 minutes later. Motorola is also working with TekNow, Inc., a specialist in paging software development, introducing a software package that allows users to send both alphanumeric material to pagers and digital data to mobile computers on paging channels. Swatch, the watchmaking unit of SMH/Swiss Corp. for Microelectronics & Watchmaking Industries, has turned out a watch with limited paging features. It's called the Piepser (German for beeper). A newer version of the watch will have a digital display.

With few exceptions, paging has not done as well internationally as it has in North America. In 1991, Motorola shipped 100,000 pagers to the People's Republic of China. They were so popular that, a year later, the company had delivered a total of 1 million pagers to China. Growth rates have been much slower in other areas of the world. In the U.K., penetration rates barely exceed 1 percent of the population compared to more than 5 percent penetration in the U.S. Hutchison Telecom, a major supplier in the U.K., is trying to give the market a much needed boost with the introduction of five Motorola-produced color pagers aimed at the untapped consumer market in the U.K. Another new Hutchison feature in the U.K. is PULSE, a pager-based financial information service for U.K. subscribers.

In Japan, Nippon Telegraph & Telephone (NTT) is offering two Mtel services, SkyPager and SkyTalk, to traveling businesspeople.

Mtel International's joint venture in Hong Kong, Sky Telecom Services Ltd., was awarded a government license for nationwide and international messaging in Hong Kong. Mtel also operates a nationwide messaging system in Mexico with its joint venture partner, Grupo Televisa S.A. de C.V. de Mexico. In Thailand, Mtel and Singapore Telecom International jointly own 40 percent of Shinawatra Paging Service, Ltd. Mtel has also been granted a license by the government of Malaysia to develop a nationwide and international messaging system operated by SkyTel Systems (Malaysia) Sdn. Bhd., a joint venture of Mtel International, Inc., and Sector Sdn. Bhd. of Malaysia. Mtel also operates Mtel U.K. Ltd., a result of Mtel's acquisition of Inter-City Paging Ltd. in 1989.

Through its ownership of Inter-City, Mtel has access to the European Radio Messaging System (ERMES), a Pan-European network protocol that is expected to be in operation in the mid-1990s. If available internationally, ERMES presumably would allow subscribers anywhere in the world to move to another area without having to obtain a new paging device. The Paging Technology Committee of Telocator in the U.S. has been working on something similar to ERMES called the Telocator Data Protocol (TDP) and believes that ERMES could greatly improve its data throughput by adopting some of the techniques embodied in its TDP. However, since ERMES is already a published standard, PCIA (Telocator) officials aren't sure how far they can go to suggest modifications to the European standard. Also, since the ERMES Working Group has disbanded, it's unlikely that the American PCIA will be able to have any influence over the emerging European standard any time soon.

Motorola has responded with the introduction of its own paging protocol called FLEX. The key to FLEX is its ability to operate at three speeds—1.6, 3.2, and 6.4 kilobits per second. This offers an easy migration path from today's 1200 bits per second (bps) with the pager following the system bit rate. Operators currently using 1200-bps technology can add FLEX 1600 bps to their system by upgrading their paging terminals while continuing to service existing pager units with relatively little investment in the infrastructure. In time, Motorola believes that paging operators will move to 3200 bps and then 6400 bps and begin to populate the market with upwardly compatible FLEX pager units—a process Motorola likens to the radio broadcast industry's shift from monophonic FM to stereo.

At maximum speed, Motorola claims that FLEX can handle 600,000 alphanumeric pagers per channel, a 300 percent increase over the capacity of the Post Office Code Standardization Advisory Group (POCSAG), the paging industry's standard. Motorola says carriers can mix traffic using POCSAG or GOLAY (another Motorola paging protocol) and FLEX on a single system. Another feature of FLEX is that it energizes the pager's electronics only when data is to be received, producing a 10-fold improvement in battery life.

Motorola is expected to begin delivering pagers that accept the FLEX protocol late in 1993. Two major Motorola competitors, NEC Corp. and Glenayre Technologies, Inc., have already licensed the FLEX technology. NEC plans to begin selling FLEX pagers in the U.S. when carriers install the infrastructure and Glenayre will produce equipment using the FLEX protocol. (NEC America and Glenayre Technologies have also informally demonstrated ERMES hardware in the U.S., and NEC Japan and its U.K. subsidiary have built about 2,000 ERMES pagers to test in Europe.) Three U.S. carriers, MobileComm, Paging Network Inc. (PageNet), and PacTel Paging, are field testing FLEX.

Motorola has also developed what it calls its Mobile Networks Integration Technology, or Monet, a series of software tools that serve as application program interfaces (APIs) between disparate wireless and wireline messaging networks. Monet will be licensed to private and public wireless/wireline data network operators, application developers, and manufacturers of computing and communications devices and infrastructure equipment. Motorola hopes to entice software producers to embed its APIs in their most popular software products. Motorola offered few details of the technology when it was introduced in early summer 1993; however it works, it has already been endorsed by RAM, ARDIS, Tandum Computers, and several other potential users.

If you were worried about product obsolescence when you bought your first camcorder, you may get a little uptight about purchasing a product that probably will work very well for a long time, but will be technically obsolete six months after you buy it. That's the name of the game in consumer electronics. Indeed, the technology is moving so fast that paging and mobile data and messaging services are beginning to blur. This is a problem, and it raises such questions as, "Will wireless messaging, including paging, compete with PCS, which is supposed to be cheaper than cellular?" Most analysts don't believe so and for a number of reasons. Telecommunications industry specialists at Prudential Securities Inc. have observed in the PCIA's "Paging Universe," a financial news brief offered by the trade organization, that the FCC's Office of Policy and Planning, in its study, "Putting It All Together: The Cost Structure of Personal Communications Services," sets the annual cost of operating a PCS network at $546 per subscriber (assuming a 10 percent household penetration). Assuming a 20 percent markup, Prudential Securities says this estimate works out to a consumer price of about $55 per subscriber per month or about $60 per month including equipment and infrastructure costs.

At the rate pager prices are dropping—an average of 5.7 percent a year since 1988—Prudential Securities estimates that by 1995 (around the time PCS is expected to become widely available) paging service will cost about $9.70 a month; that's approximately six times cheaper than the FCC's PCS price projection or about the same ratio as cellular to PCS today. For this reason, the Prudential Securities analysts say they are "not convinced that PCS will adversely affect paging any more in the future than cellular has in the past."

In fact, about a third of cellular customers are also pager users. SkyTel claims that 60 percent of its subscribers carry both cellular phones and pagers. Motorola and other pager manufacturers are already responding to this phenomenon by introducing products that integrate both cellular and pager functions. For $9.95 a month, a Sprint Corp. customer in Las Vegas can receive a short text message displayed on a cellular phone's digital screen, or be paged. The unit also will alert the subscriber to voice-mail messages, which the user can retrieve by pressing the phone's "send" button. Because the paging and voice-mail information are transmitted digitally, the service uses special Motorola equipment that digitizes the transmitted data so it can be sent over the analog cellular phone network. McCaw Cellular Communications plans to begin offering a similar service in 1994. Also, SkyTel and Motorola are market testing a new device called the MicroTAC RSVP, which combines cellular and paging service. The trial uses Motorola's AMPS and Mtel's dedicated 931.9375-MHz nationwide paging band. When the MicroTAC RSVP receives a page, the subscriber pushes the unit's "recall" button and the caller's phone number is displayed. The subscriber then pushes the "send" button to call the displayed number.

Undoubtedly, there will be new players in this market, offering new services. The National Association of Broadcasters (NAB) and the Electronic Industries Association's Consumer Electronics Group (EIA) have already formed a National Data Broadcasting Committee to develop a voluntary technical standard for high-speed data broadcasting for NTSC television stations. (NTSC is the TV broadcast standard in North America and Japan.) The two trade groups believe that a data receiver could be added to television sets or developed as a new class of consumer electronic products with outputs to fax machines, computers, or TV sets. As they see it, services may one day be offered by TV stations that include advertising-supported fax broadcast services, information/news summaries broadcast to large numbers of portable and desktop computers, and coupon or promotional information made available as a supplement to on-air advertisements.

Lotus, which expects to play a major role in the mobile computing and wireless communications, has its own vision of how the wireless future will evolve: In the short term, E-mail will drive the use of wireless. Wireless data networks will gain acceptance around the world but will not be seamlessly connected to each other. The size of external radio modems will shrink. Mid-term, say, two to three years out, nationwide all-digital networks may appear that can offer both circuit and packet-switched communications. Short wireless messaging will gain acceptance as a new form of communications and will be the compelling functionality of portable digital communicators such as so-called personal digital assistants.

Alliances between international networks will appear so that people who travel internationally will be able to communicate back to their office with ease via wireless. Radio modems will be able to support multiple networks, including both RAM and ARDIS. New applications for wireless will appear that in the past were not obvious. Long term, that is, beyond three years, Lotus expects consumers to become heavy users of wireless data technology. Many will rent public electronic mailboxes. Users will be able to communicate down the hall (WLANs), in densely populated cities (metropolitan area networks, or MANs), and in the middle of a desert (WANs)—transparently and seamlessly.

Chapter 8

The Mobile Computer Market

Can We Talk?

Listening to some industry experts talk about the emerging mobile computer market is something like listening to comedian Jackie Mason's Jewish diner routine: I would like it to be a huge market, but maybe it will just be big. It seems like a niche kind of thing, but I don't think so. It could be a gigantic thing, but who knows?

For companies like AT&T, Hewlett-Packard, Apple Computing, and Motorola, there is little doubt—it's going to be a very large market. Companies with names like Psion Plc, Itron Inc., and Telxon Corp. agree. They're all betting that laptops, palmtops, notebooks, and subnotebooks will sell at a faster rate than desktop computers, which continue to rack up double-digit percentage increases.

But who is going to spend $4,000, $2,000, or even $1,000 on a computer they can carry around just to do some of the things they would do at their desks? Probably not more than several million people, which is not a lot considering the heavy investment, high hopes, and hype that are going into this market. However, if these devices can communicate, if they can send and receive data anywhere at anytime, now we're talking about a totally new product category. This is the stuff that new consumer electronics markets are made of. VCRs hit the market priced well over $1,000; today, they're available for a few hundred dollars. But not yet.

The first generation of personal communicators haven't exactly received rave reviews and, as a result, early sales fell well short of expectations. Apple Computer's Newton MessagePad sold 50,000 units in one month when it was introduced in August 1993, but sales dropped to an average of about 7,500 a month on the heels of some tough reviews in the computer trade press and several major business publications. The nationally syndicated cartoon strip "Doonesbury" picked on the Newton for a week, focusing on its erratic handwriting recognition performance, which Apple had touted as one of its most important features. *Advertising Age* columnist Bob Garfield didn't help when he devoted an entire column to the new Apple device in September in which he concluded that "the Newton doesn't work well." *Business Week* followed with a full-page report on the Newton's "shaky" reading skills.

AT&T's EO model which costs up to $2,500 with all the options, including a cellular telephone, was selling at the rate of only about 1,000 units a month following its introduction in March 1993. Another model, the Zoomer, made by Casio and sold by Tandy Corp., wasn't doing much better.

The vision is that people will be able to communicate with each other from great distances, wirelessly, as easily as if they were sitting across the room. We're not quite there yet. But a growing legion of vendors—some of them fewer than three years old—are working very hard to make it happen with a whole new class of portable devices—personal communicators, personal information processors (PIPs), personal digital assistants (PDAs), and per-

DOONESBURY.© G. B. Trudeau. Reprinted with permission of Universal Press Syndicate.

sonal intelligent communicators (PICs)—to deliver what Microsoft's Bill Gates calls "information at your fingertips, anytime, anywhere," and what former Apple Computer CEO John Sculley refers to as "knowledge navigators."

The speed at which the computer industry is advancing the capability of these devices, including the ability to transfer data, is giving technical obsolescence a bad name. The 18-pound machine with little memory and data storage has shrunk to barely 2 pounds with 486 microprocessor power, seamless radio modem-based communications, a pen-based touch screen and/or keyboard to enter data, and a color screen to create business graphics, develop forms, manage databases, run spreadsheets, and do a lot more.

Virtually every major computer manufacturer is exploring ways to apply its technology to these products. Apple's instant success with its PowerBook is a good real-life example of why there is so much interest and activity in the mobile computers. In the first 12 months following its introduction in October 1991, Apple sold more than 400,000 PowerBook computers, generating more than $1 billion in revenues. They are selling so well that products are being developed as add-ons specifically for the PowerBook—like Applied Engineering's AErport, a telephone keypad and answering machine, and StarLAN, a wireless local-area network attachment by Teledyne Microwave, designed to sell for under $100 retail.

Market projections for personal computers are wide ranging, but the numbers are generally impressive. Dataquest, Inc., for instance, says 398,000 palmtops were shipped worldwide in 1990, mostly in Japan. By 1994, the market research firm expects sales to jump to 5.2 million units. BIS Strategic Decisions estimates that by 1995, the personal digital device category will generate $155 million in sales. AT&T Microelectronics, already a major player in this market as a components supplier, says it can only guess how big the market is going to be so it has published a small booklet, "Vision for Mobile Communications and Computing," in which it concludes that projections of related markets offer some clues as to what to expect.

For example,

- By 1995, the pen-based computer market could grow to 6 million units in the United States and could double those figures in worldwide shipments.

- Approximately 23 million cellular phones will be in use by 1996—41 million by the year 2001.
- The population of telephone subscribers will triple over the next decade, reaching 1.6 billion by the turn of the century.
- The personal computer user population will grow from about 100 million in 1990 to some 300 million by 2000.

Based on these and other related market projections, AT&T Microelectronics believes there could be 1 billion personal communicator users by the year 2000—a population that approaches the size of the television market.

AT&T also believes that "very soon," personal communicators

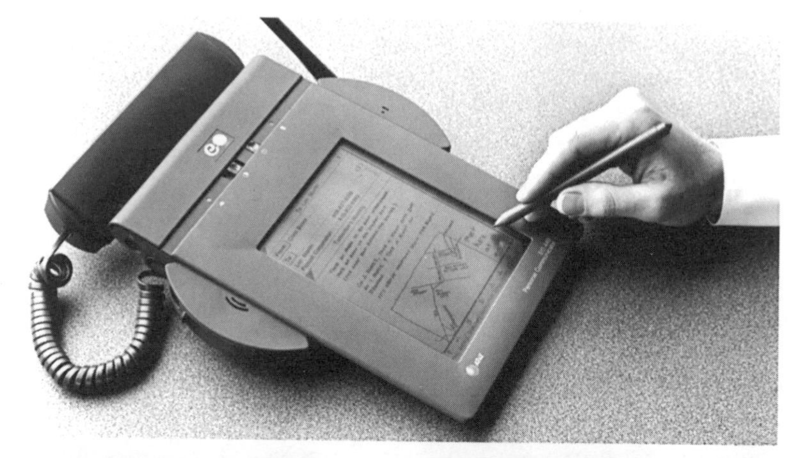

The AT&T EO 440 features wireless voice and data communications, fax, electronic mail, and pen-based computing.

will be capable of two-way paging, voice-mail, speech and handwriting recognition, analog cellular, and speakerphone functionality. Coming "sooner than you think," says AT&T, are personal communicators with such advanced features as digital cellular, Integrated Services Digital Network (ISDN), high-resolution still imaging, full-motion video, and videoconferencing. Portable video communications is a few years off, but that will come, too. Apple Computer says it's working on a version of its Newton family of PDAs that will not only communicate with other Newtons, LANs, and printers, but will be voice activated and perform as an "ultimate wireless control unit" for TVs, VCRs, and telephones.

The marriage of portable computing and communications is already being felt in all sectors of today's land-based mobile radio markets: private business-owned systems, paging systems, and cellular phones. In Motorola's view, three key factors are fueling this growth: first, the expansion of public data networks; second, the availability of small, low-cost radio modems; and, finally, the development of application and connectivity software. From the consumers point of view, the keys to success of these new products are portability and wireless connectivity.

From the manufacturers, that is, the product developers' point of view, the key to success may be getting to market as quickly as possible.

AT&T has sought to carve out a leadership position for itself by taking a majority interest in EO Inc., a California-based start-up and pioneer in the development of personal/portable computers. EO believes that more than 100 million personal communicators will be sold by the year 2000, representing $20 billion in annual sales. Motorola, certainly another key player, has joined with AT&T, Apple, Sony, Matsushita, and Philips to offer a wireless PIC based on General Magic's Magic Cap, an object-oriented operating system, and Telescript, the programming language and communications software package that General Magic is promoting as a de facto industry standard. Telescript allows different networks and applications to work together and lets users customize the software's functionality. AT&T also plans to offer a "telephone network version" of the 3DO communications-capable portable computer, probably by mid-1994.

Initially, Motorola will produce and market its PIC with fully integrated, two-way wireless communications functionality for the business market; the mass market will come later. This will be the first of a number of personal communications products to be introduced by Motorola's new paging and wireless data group, such as PIC versions that will operate on several wireless data networks, including ARDIS and RAM Mobile Data, as well as

Simon™, designed by IBM and marketed exclusively by BellSouth, integrates a cellular phone, wireless facsimile machine (send/receive), pager, electronic mail, calendar, appointment scheduler, address book, calculator, and pen-based note pad/ sketchpad.

with "emerging carriers." Like Apple and its PowerBook, Motorola is also working with third-party software developers and peripheral manufacturers to develop a variety of end-user applications.

Hewlett-Packard, which has had tremendous success with its HP95LX palmtop, has since introduced a follow-up product that is optimized for the mobile user. The new HP100LX is DOS based and preloaded with Lotus Development Corp.'s cc:Mail, the most widely used E-mail software. It weighs 11 ounces and has two Type 2 PCMCIA-compatible card slots. Even newer is HP's OmniBook 300, a subnotebook computer designed to operate wirelessly with personal computers, networks, and printers. At just under 3 pounds, with a 9-inch screen and a keyboard with standard-sized keys, the OmniBook runs the latest version of Windows. It has four slots for PCMCIA cards, or room for miniature hard disk drives, modems, or other devices. And it will run for four to nine hours on a charge of its nickel-metal-hydride batteries or with 4 AA batteries. But it's not cheap: The 40M-byte disk drive–equipped HP OmniBook is priced at $1,950; a 10M-byte flash disk version is $2,375. The fax/modem communications pack and printer port are extra.

BellSouth Cellular has unveiled a fully integrated, IBM-designed personal communicator with a cellular phone, wireless fax (send/receive), pager, E-mail, calendar, appointment scheduler, address book, calculator, and pen-based note/sketch pad. IBM has given BellSouth exclusive distribution rights to the unit in the United States. Called Simon and priced under $1,000, the device's PCMCIA slot allows users to add a paging card to receive messages on a national, regional, or local basis, through MobileComm, BellSouth's paging company. The slot can also be used to upgrade or expand the memory for increased storage capacity. Simon weighs just over a pound and uses an LCD display as a keypad and touch screen.

GRiD Systems Corp.'s newest PalmPad is another example of the technology available to consumers. The GRiD unit integrates communications in a 2.9-pound computer, and has an optional RF module that allows it to operate with either ARDIS or RAM Mobile Data networks. GRiD plans to expand the PalmPad's functionality with a new model with at least two Type 2 PCMCIA slots.

Compaq Computer Corp., another well-known name in portable computing, will introduce a hand-held, pen-based computer in early 1994. This unit will weigh less than 2 pounds and will be priced from $1,000 to $2,000. But unlike Apple's PDA, which is aimed primarily at the consumer electronics market, Compaq's "Mobile Companion" will be targeted at business users.

Sun Microsystems, Inc., the largest producer of computer workstations for scientists and engineers, is a total newcomer to this market. Sun has formed a mobile computing subsidiary called FirstPerson Inc. and says it is developing a new class of portable computer for the consumer electronics market in direct competition with General Magic and its supporters. But Sun doesn't expect to begin selling its first portable product until at least 1995.

Like EO, General Magic, and Apple Computer, Sun is trying to line up support for its proprietary technology. Initially, the company's focus will be on developing software that will allow different digital devices to share data and actually work together, which is also part of General Magic's product development game plan. It has licensed designs for prototype models of at least two mobile computers to two Japanese companies, and it has signed a licensing and joint development agreement with a private Russian corporation called Elvis+ Ltd. to advance wireless communications technologies. Sun is also trying to line up support in Europe.

Other pen-based mobile computers have been introduced or announced by NCR, Casio, Tandy, Fujitsu, Zenith Data Systems, and Sharp Electronics. The Tandy and Casio pocket-size Zoomer will perform notebook/calendar/to-do list/address book–type functions with the help of a built-in alarm and world-time clock. It also has a full-function calculator; personal financial management with Pocket Quicken software; a built-in dictionary, spell checker, and thesaurus; and it translates words into 26 different languages. You can write on the screen and let the system translate your letters into type, or you can call up a picture of a keyboard and tap the letters of the keyboard with a stylus. It also downloads data from a variety of PC software and can send and receive E-mail with other modem-equipped PCs, including other Zoomers.

AT&T has two partners in EO, both Japanese—Matsushita and Marubeni. AT&T's agreement with EO allows the smaller company to include the AT&T brand on its personal communicators and gives EO access to AT&T's communications and component technologies and services, as well as to its sales channels. Matsushita's role is as a low-cost, high-volume manufacturer and component supplier. Marubeni, a major trading company with some 200 offices worldwide, will handle global sourcing duties and help EO gain access to foreign markets.

EO introduced its first products in June 1993, the AT&T EO 440 and 880. Users of the 2.2-pound device may subscribe to AT&T EasyLink Services' AT&T Mail electronic mail, which also features a built-in microphone and speaker for voice annotation of forms, documents, meeting notes, and calendar appointments. An optional cellular module provides both voice and data communication over a wireless network and an industry-standard Type 2 PCMCIA slot. The 4-pound EO 880, which has two PCMCIA slots, features a port for connecting a full-sized VGA monitor for presentations and a SCSI II port for connecting external hard drives. But, at an introductory price of nearly $4,000

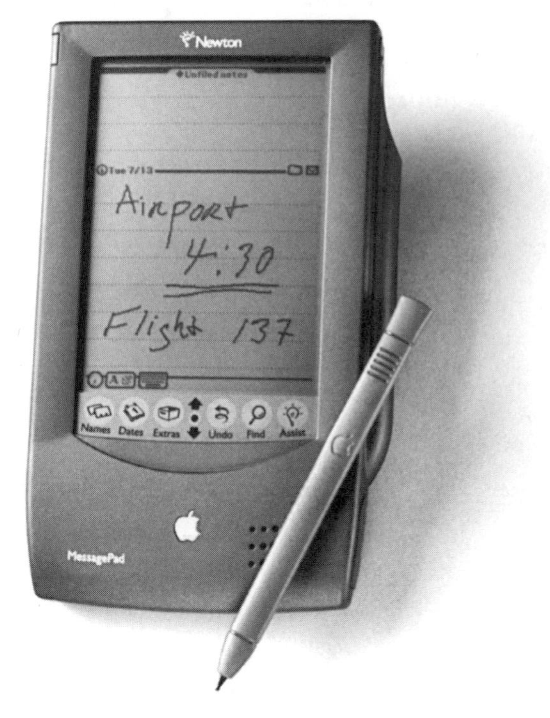

Apple Computer's Newton™ MessagePad™ received some negative press when it was introduced because of technical problems, but it is one of the best selling "smart assistants" on the market.

for a full-featured model, the EO devices are not likely to be mass market items, which is why EO is working on a smaller, more highly integrated (the phone functions are built in) device.

What's the difference between today's personal communicator/computers and computer/communicators? For one thing, the computers usually communicate through optional or add-on devices. That will change as notebook computer vendors seek to enhance their products. To a lesser extent, the same is true of electronic organizers, such as the Sharp Wizard, which probably will add paging, or a more advanced form of wireless messaging. Products will be available to accommodate every user's needs. It will be increasingly difficult to tell the difference between PDAs, PICs, and similar devices. But manufacturers are already attempting to make those important definitions. In true competitive fashion, EO says it views Apple's Newton MessageCard more as a pocket organizer than as an "anywhere, anytime" communicator with the capabilities embodied in the EO product. That's because, with the exception of a built-in infrared "beam" for sending messages across a 1-meter distance, the initial version of the MessageCard can only send E-mail messages via an external modem. Now, however, Newton MessagePad users can receive wireless messages from a variety of E-mail sources through a service provided by RadioMail Corp. Users communicate with Newtons by sending their messages to RadioMail via the Internet. RadioMail relays the message via the MobileComm paging service to any Newton equipped with a messaging card in its PCMCIA slot. RadioMail expects to begin offering two-way links for Newton MessagePad users by mid-1994.

In time, the MessagePad will be able to access a wireless messaging service both nationally and locally. Apple has already signed on with SkyTel to support most of Apple's wireless communications needs. "Telecommunications," says John Sculley, "will be as important to Newton as graphics are to Macintosh."

In an effort to make its Newton technology a pervasive standard available to the computer, consumer, and telecommunications industries, Apple has offered to license key technologies to companies who will produce and market Newton-like products. Several companies have already committed to the Newton concept:

- Sharp Corp. will help develop and manufacture Newton, and is a licensee, along with VLSI Technology and GEC-Plessey Semiconductor, of the ARM chip—the heart of the Newton product family. (ARM processors will also be used in the next generation of video games.)
- Motorola will make and market a hand-held device featuring integrated, wireless communications capabilities based on the Newton operating system.
- Siemens Private Communications Systems Group and ROLM, its U.S. subsidiary, will collaborate with Apple in the development of NotePhone, a combination of Siemens-ROLM telephony and Newton technology designed to provide access to telephone and fax features.
- Cirrus Logic will support Newton technology by developing and supplying Newton-compatible chipsets for use by licensees of the Newton operating system and by Apple itself in its planned Newton products.
- Kyushu Matsushita Electric (KME) has agreed to license the Newton operating system from Apple for use in future products. Apple and KME also intend to explore opportunities for using KME-provided technology in future Newton family products.
- LSI Logic will manufacture an application-specific integrated circuit (ASIC) chip for Newton PDAs. The custom ASIC provides the interface between the microprocessor, the memory, and the user interface.

Apple has also formed partnerships with three regional phone companies to provide communications service for Newton PDAs. Apple and BellSouth plan to test a display phone that would provide banking, fax, messaging, and other home information services. Ameritech will assist Apple in fax and voice-messaging, and US WEST will work with Apple to develop screen-based telephone services.

Motorola's licensing agreement with Apple apparently will not affect its alliance with General Magic. Motorola's goal is to support all major industry operating platforms, as well as all wireless networks, but Motorola will market its first personal communicator device based on the General Magic platform.

Fujitsu Personal Systems, Inc., which marketed one of the first portable pen-based computers, also is working with GO. The two companies will offer a Developers Dream Kit for pen computing applications. The kit will help software developers integrate Fujitsu's 325Point pen-tablet with pen software from GO, Slate Corp., and PenMagic.

Obviously, software will play an increasingly critical role in the mobile computing market. Products like Telescript, Magic Cap, and the Developers Dream Kit and software developers like Palmtop, Lexicus, and Paragraph International will continue to be vital elements in the development of wireless portable computing, especially if they hope to be able to differentiate their products.

AT&T is demonstrating how that can be done. It is working with two software developers, GO and PennStuff, and Trimble Navigation, which makes and markets satellite-based global positioning systems (GPS), to make it easier for other software developers to turn their personal communicators into "personal navigators." Users will be able to find their location on maps, navigate point to point, find restaurants or businesses on the go, and even call up current traffic and weather information. The four partners have demonstrated their GPS application programming interface (API) on an AT&T EO 440. A software development kit includes a sample source code developed by PennStuff and uses a Trimble PCMCIA GPS sensor.

AT&T and GO have also attempted, with the help of *Pen Magazine*, to do some blue-sky product planning. One example is the Office PhonePad, a personal communicator with a telephone base and detachable portable computer. Another is the Compact PC, a full-function pen-based computer/communicator with built-in makeup kit for the busy woman traveler, available in a variety of designer colors. A third proposal is the CellPad, a portable cellular phone/computer notepad with digital sound processing to store and schedule outgoing voice messages, an answering machine, and a digital pager.

Sony's long-range personal communicator plans are even more ambitious: By the turn of the century, the consumer electronics leader hopes to introduce a portable product that not only communicates globally (and, of course, wirelessly) but also translates languages automatically and in real time.

For the most part, personal communicators will likely follow the pattern set by cellular phones. Mobile, high-productivity professionals will be the "early adapters," followed by mainstream consumers. Specially designed units will be available for heavy-use vertical applications such as sales automation, insurance adjustment, and field data entry and retrieval.

There is little doubt that this is going to be a huge market. "A lot of technology is searching for a marketplace," says BIS Strategic Decisions. "But vendors need to have a better understanding of users' needs before deciding how to sell these devices." That may take some time, but it will come. EO estimates that 100 million personal communicators will be in use by the end of the century. If that's true, personal communicators could outnumber desktop personal computers.

CHAPTER 9

UNHEALTHY PUBLICITY FOR CELLULAR PHONES

For a long time, it was high-tension wires. Then, television sets, microwave ovens, electric blankets, electric shavers, and hair dryers. More recently, it was police radar. Now, it's cellular phones.

The unhealthy publicity for the cellular industry started in January 1993 when a Florida widower claimed on the "Larry King Live" television talk show on CNN that his wife died from a brain tumor caused by her cellular phone. That show, his lawsuit against a cellular phone manufacturer and retailer, and the media blitz that followed, threw a big scare into cellular phone users—and the cellular industry.

Cellular equipment manufacturers, carriers, and their trade association responded immediately, pointing to the hundreds of studies made over 40 years about the potential health risks from radio waves; none of these studies have turned up any solid evidence of human brain damage. But journalists and Wall Street analysts weren't convinced. Cellular stocks began to fall in heavy trading. Within days, Motorola's stock dropped by 17 percent. Some cellular subscribers canceled their service; others simply decided not to buy a cellular phone. Most consumers sat tight. In one California market, where McCaw Cellular Communications has a customer base of about 90,000, only 20 calls were received in response to the health scare stories.

Almost immediately following the Larry King show, the Cellular Telecommunications Industry Association (CTIA) began nightly telephone polling to track public opinion. With national press coverage and several days of interviews with "experts" on the morning network TV shows, public awareness was very high. Only a small number were convinced that cellular phones cause cancer. A lot more dismissed the whole thing as a sensation-seeking media making more of the story than was there. In the middle, nearly half the public indicated concern, but weren't sure what to believe.

The industry decided to fund additional research on cellular safety, and it asked the Food and Drug Administration (FDA), the Federal Communications Commission (FCC), and the Environmental Protection Agency (EPA) to appoint a blue-ribbon panel to direct new research into cellular phones—research that would be credible to the public. "It is truly amazing," an angry CTIA President Tom Wheeler told industry executives at the association's annual meeting in Dallas shortly after the crunch of national publicity, "how unsubstantiated allegations in one lawsuit can be hyped by the media to the point where, almost overnight, a high percentage of Americans are aware of the specious claim and the stock market takes a nosedive." No one, Wheeler stressed, has demonstrated any link between cellular and cancer.

Cellular system operators attempted to calm customers' fears with packets of information mailed with bills. The CTIA distributed follow-up material to the media, noting, for example, that portable cellular phones operate at a maximum of only 6/10 of a watt of power, approximately 15 percent below the FCC's safety threshold. In a typical urban market sampled by the CTIA, cellular phones operate at full power only 5 percent of the time. The rest of the time they operate at a power level of less than half the maximum 6/10 of a watt. Further, in its defense, the CTIA pointed to a study of actual usage records in a sample metropolitan area, noting that 71 percent of all cellular calls last less than 90 seconds, while 85 percent of cellular calls last less than 3 minutes, not nearly long enough to cause any physical damage to a cellular phone user.

Representatives of the FDA, EPA, and National Cancer Institute all testified at a briefing called by the House Telecommunications Subcommittee that there is no conclusive evidence linking cellular phones to brain cancer. In fact, the National Cancer

Institute said that among people under the age of 65—the target population of cellular users—there had been no increase in brain cancer and no change in the randomness of the tumors' locations.

Articles appeared for weeks, but not all the news was bad. Some journalists called Bellcore, the research arm of the seven regional Baby Bell telephone operating companies, to ask about its 1992 study of local telephone company workers' exposure to electromagnetic fields (EMF). Results of the study showed workers' exposure to EMF was almost as low on the job as in their own homes. In fact, telephone company employees who spend much of their day in a variety of work environments are exposed to approximately the same EMF levels as employees measured away from work.

Part of the problem is that most research in the field has focused on low-frequency emissions, such as those produced by overhead power lines. Very little work has been done at the high-frequency end of the spectrum where cellular phones operate. The most recent review of EMF research was completed in 1991 by a 120-person panel of experts assembled by the Institute of Electrical and Electronics Engineers (IEEE), the largest technical society in the world with more than 300,000 members. Of the 321 studies reviewed by that panel, 29 dealt with cancer and none of those included human experiments. The IEEE Committee on Man and Radiation's official position is that, based on present knowledge, prolonged exposure to RF fields from portable and mobile telephone devices at or below the recommended levels is not hazardous to human health.

The scientifically accepted method for establishing safety standards is to identify levels at which problems occur and then reduce exposures below those levels. In the case of portable cellular telephones, the operating levels are significantly below those at which safe operation has been demonstrated, according to "Biological Effects of Radio Frequency Radiation," a 1984 EPA review of the potential of radio waves to induce cancer.

Towers constructed for cellular base stations have also come under attack, but usually only locally. Nevertheless, community residents have been very effective in suppressing attempts by cellular operators to erect cellular towers in residential neighborhoods. Typical of recent efforts is the research conducted by a group calling themselves Citizens Against the Tower in Bedminster, New Jersey, which included wading through reports on the earliest research and petitions filed with the FCC. The group produced enough data to convince the town council to vote down a proposal by Cellular One to erect a 110-foot monopole, on which nine antennas were to be attached.

From a legal standpoint, the FCC has a limited role in electromagnetic radiation matters. But it does have an obligation under the 1969 National Environmental Policy Act to ensure that its licensing and regulatory actions do not create adverse health or environmental effects. Since the health scare broke, the FCC has proposed updating its safety standards concerning electromagnetic radiation (EMR) exposure to those adopted in 1992 by the American National Standards Institute (ANSI) and the IEEE. That standard exempts cellular transceivers operating at a frequency of 850 MHz with power levels below 0.74 to 0.76 watts. (Cellular phones currently operate at 0.6 watts.)

Applying the same standard to the next generation of cellular systems, power levels would have to be between 0.31 and 0.39 watts to meet ANSI/IEEE specifications. The FCC has also authorized total effective radiated power levels up to 500 watts for cellular towers, depending on geographic area and tower height. Most cellular towers in large cities operate at about 10 watts per channel. The Telecommunications Industry Association (TIA), which represents equipment manufacturers and works with other trade organizations to develop industry technical standards, is encouraging the government to put into law what is already the de facto standard for land mobile manufacturers.

The problem is not unique to the U.S. The European Space Agency is investigating the potential for health problems from using portable phones with mobile satellite systems. Sweden, a country with more than 12 years of commercial cellular experience and one of the highest market penetration levels in the world, has found no evidence that cellular phones cause cancer. Specifically, research conducted at Sweden's Lund University has found that EMR does not promote cancer from nonthermal exposure levels at 915 MHz, the range in which some cellular phones operate in that country. Abstracts of technical papers presented before the European Bioelectromagnetics Association early in 1993 indicate that, "To date, no convincing laboratory evidence has been obtained indicating that EMR causes tumor promotion at non-thermal exposure levels." But Sweden continues to study the issue and is considering placing certain restrictions on products that generate EMR. Sweden already has restrictions on how close people can live to high-power lines.

Actually, the entire wireless personal communications industry has been pretty much up front about the issue. In early 1991, in its petition to the FCC for spectrum for Data-PCS, Apple Computer specifically asked the commission to consider the health consequences in making its decision. Proxim covers the issue at length in a pamphlet for resellers of its RangeLAN wireless local-area network. Noting that its products operate in the 902–928-MHz range (the spread-spectrum frequency, adjacent to the cellular phone band), Proxim points out that the FCC limits the power output of spread-spectrum devices to 1 W (1000 mW). Most vendors' products operate considerably below this level, typically in the 100–250-mW range. A cellular car phone antenna has an output power of 3 W (3000 mW), which is about 10 times higher than most WLAN products, but still within acceptable limits set by ANSI, the IEEE, and now, the FCC.

Ironically, in 1986, the EPA terminated an in-house EMR research program that may have answered most of the questions concerning some cellular phone users today. It turned out to be the last federally funded civilian research program in the field. Concerned industry leaders and officials of the FCC and National Telecommunications and Information Administration (NTIA) tried to convince the EPA to set RF exposure guidelines, but the EPA dropped this initiative in 1988, claiming budget problems and other priorities. That's not likely to happen again. "There's a level of sophistication in the questions being asked by consumers calling the FDA," Dr. Mays Swicord, the head of the FDA's Radiation Biology Branch, told a CTIA audience during the association's 1993 conference. "We cannot brush these off. We need to explain levels of uncertainty and risk."

Herschel Shosteck Associates, Inc., a cellular industry consultant, has counseled cellular phone retailers who are asked, "Does cellular cause cancer?" to reply: "It's very unlikely. However, no one knows for sure. It's for this reason that the Cellular Telecommunications Industry Association has asked the federal government to oversee a full restudy." The dust seems to have settled. But as the CTIA's Tom Wheeler told association members at their meeting in Dallas, "the wireless future is not just new services but new realities of perception."

As might be expected, some people are trying to benefit from all the negative publicity surrounding the cellular health scare. Several small, entrepreneurial companies have introduced devices claiming to shield cellular phone users from any harmful radiation. Dynaspek has introduced Cell Shield, a $30 device that fits over a cellular phone's antenna; Dynaspek claims that its product reduces radiation up to 95 percent. Quantum Laboratories has come out with Cellguard, a $50 unit that covers both the antenna and earpiece and that it promotes as "a protective shield that reduces the risk of cellular phone use." Another device, called Cellularshield, claims to either block or redirect most electromagnetic radiation from a cellular phone's antenna. Considering how quickly American entrepreneurs respond to a market opportunity when they see one, the introduction of many more such products is likely.

Hoping to put the issue to rest once and for all, the CTIA said it will spend more than $1 million on new EMR research and per-

Tests conducted for the National Institutes of Health (NIH) by Dr. Om Gandhi, professor and chairman of the Department of Electrical Engineering at the University of Utah, indicate that radiofrequency (RF) exposure from hand-held cellular telephones is well within national safety standards.

haps an equal amount on public information programs over the next five years. Independent of the CTIA, McCaw Cellular Communications, the leading cellular carrier in the U.S., hired Om Gandhi, chairman of the electrical engineering department at the University of Utah and a former AT&T Bell Labs scientist who specialized in EMR research, to conduct additional studies in the frequencies and at the power levels used by cellular phones.

In July 1993, six months after the Larry King show, the CTIA released a progress report. The association said that after reviewing thousands of studies, including hundreds that specifically address cellular frequencies, there is nothing to suggest any link between using portable cellular phones and brain cancer. The CTIA's research program is being directed by public health epidemiologist Dr. George Carlo, adjunct professor at George Washington University Medical School and chairman of Health and Environmental Sciences Group, a health research firm based in Washington, D.C. To perform this work, the CTIA has awarded grants to the School of Public Health at the University of California at Berkeley and the University of Alabama at Birmingham.

The cellular industry got some good news in August 1993 when a federal judge in Chicago dismissed a class-action lawsuit against a group of cellular carriers that alleged that cellular phones were a health threat to users. The judge said he was throwing out the case because the FCC and FDA have primary jurisdiction over the issues raised in the case, and because there was no "hard evidence" linking the use of portable phones to cancer, according to information on the case issued by the CTIA. However, the ruling has no impact on the lawsuit filed in Florida by the man who claimed his wife died from cancer caused by using a cellular phone.

Chapter 10

Future Talk—An Update

You can always tell when a new market is going to be big. Long before companies are able to figure out how to take advantage of it, others are already profiting from it with seminars, market research, executive search, financial and educational services, books, magazines, and newsletters.

Wireless communications is going to be that kind of market. It has become virtually impossible to keep up with the almost daily announcements of new technological breakthroughs, new corporate formations, changing market and regulatory trends, and product introductions and upgrades. Since the early chapters of this book were written, the FCC has removed the major barriers to the advancement of mobile communication and computing services by allocating spectrum for emerging personal communications services (PCS), AT&T Bell Laboratories has received a patent (patent no. 5,239,521) for a two-way wrist radio that would qualify the fictional Dick Tracy model for a place in the archives of the Smithsonian, and AST Research Inc. announced—months after it introduced the product, and weeks in advance of its shipment to dealers—that it was upgrading its Zoomer personal communicator.

Still, many questions remain.

Will PCS be so much better than cellular?

Clearly, if PCS is going to succeed on its own, consumers will demand much higher performance from PCS than cellular and other wireless personal communications services. That means a highly reliable and flexible system with smaller phones, improved battery life and range, and lower cost.

Cellular carriers are already offering PCS-like services in some areas allowing subscribers to use a single personal phone number to reach anyone virtually anywhere. Cellular phones themselves are getting smaller all the time, and the emergence of seamless national networks will give them just about all the range most subscribers will need.

In fact, the cellular industry in the United States is adding an average 9,500 new subscribers a day. Of course, there is something called "churn," which results in the loss to the cellular companies of up to 30 percent of their customers every year because of high airtime costs or some other dissatisfactions with the service—or subscribers simply discover they don't need a cellular phone after all. Overall, however, costs are dropping, the service is improving, and business is booming for both cellular carriers and equipment manufacturers. (Ericsson predicted in the third quarter of 1993 that surging sales of cellular phones and network equipment would double its pretax profit for the year.)

Will PCS actually be cheaper than cellular?

Few analysts are convinced that it will be cheaper. Cellular phone and airtime prices are dropping; a U.S. General Accounting Office study indicates that rates for cellular airtime declined 27 percent from 1985 to 1991, and that the average monthly bill dropped from $96.83 in 1987 to $68.68 in 1992. Some carriers are even giving phones away. McCaw Cellular has already reduced its prices by 20 percent for users of its new digital service between Florida and the Pacific Northwest. NYNEX has cut its cellular phone prices in half and offers two months of basic service for $29.95 a month with no activation fee—a savings of $50.

Southwestern Bell Mobile Systems has cut the price its cellular customers, and ultimately all cellular customers, pay for calls when using their phones away from home systems. Working with Southwestern Bell, cellular companies have agreed to eliminate roaming charges, which can be $3 or more a day, thus virtually eliminating its most often-heard complaint from customers. Meanwhile, cellular is beginning to look more and more like plain old telephone service with 911 emergency calling, weather, sports information, and similar features.

Competing with cellular is only a part of the problem facing would-be PCS providers. First, they have to build their networks and that's going to be very expensive. Starting with the cost of acquiring a license through auction, which could be as much as $2 billion, PCS operators must then string thousands of transmitters every 500 to 1,000 feet to create a viable network. A study by Arthur D. Little indicates that a network serving 7 million subscribers in a 300-square-mile area would cost about $6.3 billion.

Will pagers compete with portable computers and personal digital assistants (PDAs)?

No doubt they will for consumers with basic messaging requirements. The newest pager models can carry text messages on their screens, including news and sports information. And whereas pagers are being used to screen calls by cellular phone subscribers, pagers now come with their own priority call screening feature that essentially performs the same function.

Will mobile satellite communication services offer enough new services at low enough prices to supplant cellular?

Initially, services such as Motorola's Iridium and Inmarsat's Project 21, will be too costly for even the most ardent cellular user. The Iridium people have set some aggressive price targets, but by the time Iridium gets off the ground, it will have to compete with cellular, ESMRs, PCS, and digital, satellite-based paging. Motorola and Inmarsat are also about to be scooped by Orbital Sciences Corp. and its Orbital Communications subsidiary. Orbcomm says it plans to launch the first two 87-pound low-earth-orbit (LEO) satellites in the first quarter of 1994 and begin service a few months later.

The commercial airlines are making good progress in the implementation of new passenger communication services. Claircom Communications, the company owned jointly by McCaw Cellular and Hughes Network Systems, has received a $117 million contract to provide 200,000 seatback and armrest-mounted telephone units, as well as the airborne radio systems, to serve Claircom's airline customers. Mercury Communications, a division of Cable & Wireless Plc, has been granted licenses in the United Kingdom enabling it to launch the trial of its digital terrestrial flight telephone system, Mercury FlightLink, based on a service developed by Mercury and In-Flight Phone International. United Airlines will equip its entire U.S. fleet—more than 500 aircraft—with GTE Airfone's new, digital telecom system. SkyPhone, a consortium comprising BT, Singapore Telecom, and Norwegian Telecom, is providing worldwide airborne satellite communications services to Air India and Emirates, the international airline of the United Arab Emirates.

Will mobile data networks be able to differentiate their services from the new and emerging digital cellular systems? Or from each other?

That seems less likely as RAM Mobile Data and ARDIS attempt to make their networks interoperable to compete better with emerging services.

Will ESMR services have an impact on the cellular or PCS markets?

No doubt they will as they begin to roll out across the country. NEXTEL Communications has already activated its digital mobile network in southern California, and others aren't far behind. CenCall Communications expects to introduce enhanced specialized mobile radio (ESMR) service in the Pacific Northwest and Rocky Mountain regions, and Advanced Radio Communications Services of Florida will begin operating its ESMR in southern Florida by late 1994. Members of the financial community are very high on SMRs and believe they can compete with cellular and PCS. But the ESMRs may have a tough time competing on price over the long term, and they are going to have to do a much better job of educating the public about their services.

Are cellular phones dangerous to our health?

No one seems to know for sure. At least, not yet. The CTIA has been criticized for being too optimistic about the outcome of research being conducted in the field of electromagnetic radiation, research that should help answer questions about the safety of cellular and other types of personal communications devices. Meanwhile, the CTIA-sponsored Scientific Advisory Group on Cellular Telephone Research has been meeting with officials of the Center for Devices and Radiological Health of the Food and Drug Administration and have scheduled a series of seminars with government scientists and health experts from academia and the private sector to discuss such topics as the electromagnetic spectrum, RF exposure from cellular phones, the effects of modulation, cause and effect criteria, and cancer mechanisms. Ironically, as that debate rages on, a survey sponsored by Motorola has found that 82 percent of the nation's cellular phone users believe their cellular phones make their workday less stressful.

What does it all mean?

A lot of people, especially engineers who work in the field, like to point out that "wireless" has been around for a long time. Indeed, it has. Marconi began experimenting with the wireless radiotelegraph in 1894—100 years ago. Of course, no one would argue with the huge improvements over the years in functionality, portability, and even cost that makes the new generation of wireless products significantly more useful than anything before them. And more marketable. Like "office machines" (calculators) and professional video equipment (camcorders), wireless transcends commercial electronics into the consumer electronic arena. Former FCC Commissioner Ervin S. Duggan, now the head of PBS, says he's convinced that if *The Graduate* were being filmed today, the lead character, played by Dustin Hoffman, would be advised to get into "wireless" instead of "plastics." The smart money, says Duggan, sees wireless as the technology, and the business, of the future. Craig

McCaw, of McCaw Cellular, puts it another way: “We have shown customers the future, and they like it.”

Glossary of Wireless Terms and Acronyms

"A" Carrier: The nonwireline cellular company, that operates in radio frequencies from 824 to 849 MHz.

Airtime: Time spent on a cellular phone, which is usually billed to the subscriber on a per-minute basis.

AMPS: Advanced mobile phone service, the standard for analog cellular telephones.

Analog: The traditional method of transmitting voice signals where the radio wave is based on electrical impulses, which occur when speaking into the phone. Most cellular companies today transmit in analog.

"B" Carrier: The wireline cellular carrier, usually the local telephone company, that operates on the frequencies 869 to 894 MHz.

Base Station: The fixed transmitter/receiver device with which a mobile radio transceiver establishes a communication link to gain access to the public-switched telephone network.

Cell: The geographic area served by a single low-power transmitter/receiver. A cellular system's service area is divided into multiple "cells."

Channel: The width of the spectrum band taken up by a radio signal, usually measured in kilohertz (kHz). Most analog cellular phones use 30-kHz channels. Motorola's narrow AMPS uses a 10-kHz channel.

Code-Division Multiple-Access (CDMA): A digital technology that uses a low-power signal "spread" across a wide bandwidth. With CDMA, a phone call is assigned a code instead of a certain frequency. Using the identifying code and a low-power signal, a large number of callers can use the same group of channels. Some estimates indicate CDMA's capacity increase over analog may be as much as 20 to 1. The Telecommunications Industry Association (TIA) has awarded CDMA interim standard approval (IS-95).

CT-1: Cordless telephone—first generation, or any variety of North American, European, and Japanese analog cordless telephone.

CT-2: Cordless telephone—second generation, a digital cordless telephone standard; generally used in a residential cordless phone, a telepoint application, or a small-office WPBX system.

DCS 1800: Digital communication service at 1800 MHz. An extension of the global system for mobile communications (GSM).

DECT: Digital European cordless telecommunications. A digital cordless telecommunications system intended initially for WPBX applications, later to be used in the home market. DECT supports both voice and data communications.

Digital Modulation: A method of transmitting a human voice using the computer's binary code, 0's and 1's. Digital transmission offers a cleaner signal than analog technology. Cellular systems providing digital transmission are currently in operation in several locations for both trial and commercial service.

Dual-Mode Phone: A phone that operates on both analog and digital networks.

ETSI: European Telecommunications Standards Institute. One of the European organizations responsible for establishing common, industrywide standards for telecommunications.

FCC: Federal Communications Commission. The U.S. government agency responsible for allocation of radio spectrum for communication services.

Frequency Reuse: Because of their low-power, radio frequencies assigned to one channel in a cellular system are limited to the boundaries of a single cell. Therefore, the carrier is free to reuse the frequencies again in other cells in the system without causing interference.

GHz: Gigahertz (billions of hertz).

GSM: Global system for mobile communications (originally called the Groupe Speciale Mobile). It is the digital cellular standard for Europe.

Handoff: Cellular systems are designed so that a phone call can be initiated while driving in one cell and continued no matter how many cells are driven through. The transfer to a new cell, known as a handoff, is designed to be transparent to the cellular phone user. During a cellular conversation, when the user reaches the edge of the service area of a cell, computers in the network assign another tower in the next cell to provide the phone with continuing service.

Hertz: The unit of measuring frequency signals (one cycle per second).

ISDN: Integrated services digital network. A switched network providing end-to-end digital connectivity for simultaneous transmission of voice and data over multiplexed communications channels.

IS-54: Interim Standard Number 54, the dual-mode (analog and digital) cellular standard in North America. In the analog mode, IS-54 conforms to the AMPS standard.

ISM: Industrial, Scientific, and Medical. It is the unlicensed radio band in North America and some European countries. It is also referred to as Part 15.247, the FCC regulation that defines the parameters for use of the ISM bands in the United States, including power output, spread-spectrum, and noninterference.

KHz: Kilohertz (thousands of hertz).

Metropolitan Statistical Area: An MSA denotes one of the 306 largest urban population markets as designated by the U.S. government. Two cellular operators are licensed in each MSA.

MHz: Megahertz (millions of hertz).

Mobile Telephone Switching Office (MTSO): The MTSO is the central computer that connects a cellular phone call to the public telephone network. The MTSO controls the entire system's operations, including monitoring calls, billing, and handoffs.

PCS: A loosely defined future ubiquitous telecommunications service that will allow "anytime, anywhere" voice and data communication with personal communications devices.

PHP: Personal Handy Phone. It is Japan's standard for digital cordless telephones.

POPS: Short for "population." If the coverage area of a cellular carrier includes a population base of 1 million people, it is said to have 1 million POPS. The financial community uses the number of potential users as a measuring stick to value cellular carriers.

PTT: Post, Telephone, and Telecommunications Administration. European government organizations responsible for mail and telecommunications services within their respective countries.

Reseller: A middleman who buys blocks of cellular time at discounted wholesale rates and then resells them at retail prices.

Roaming: Using a cellular phone in a city other than the one in which you live.

Rural Service Area: The FCC divided the less populated areas of the country into 428 RSAs and licensed two service providers per RSA.

SMR: Specialized Mobile Radio. A private, business service using mobile radiotelephones and base stations communicating via the public phone network.

Spectrum: The complete range of electromagnetic waves, which can be transmitted by natural sources such as the sun and man-made devices, such as cellular phones. Electromagnetic waves vary in length and therefore have different characteristics. Longer waves in the low-frequency range can be used for communications, while shorter waves of high frequency show up as light. Spectrum with even shorter wavelengths and higher frequencies are used in X rays.

Spread-Spectrum: Originally developed by the military, spread-spectrum radio transmission essentially "spreads" a radio signal over a very wide frequency band to make it difficult to intercept and difficult to jam.

Telepoint: A cordless telephone system in which a subscriber can make but not receive phone calls in public areas which have been equipped with telepoint base stations. The system is not mobile; the user must remain essentially in a fixed location throughout the duration of the call. Both service and equipment are less expensive than cellular.

TIA : Telecommunications Industry Association. The North American organization established to provide industrywide standards for telecommunications equipment.

Time Division Multiple Access: The cellular industry established a TDMA digital standard in 1989. TDMA increases the channel capacity by chopping the signal into pieces and assigning each one to a different time slot. Current technology divides the channel into three time slots, each lasting a fraction of a second, so a single channel can be used to handle three simultaneous calls.

WLAN: Wireless local area network. A computer network that allows the transfer of data and the ability to share resources, such as printers, without the need to physically connect each node with wires. WLANs may also offer mobility within an office or similar environment.

WPBX: Wireless private branch exchange. The WPBX offers business users the ability to make and receive calls using cordless telephones anywhere on a company premise.

Sources: Cellular Telecommunications Industry Association (CTIA), National Semiconductor, and InterDigital Communications Corp.

Directory of Wireless Personal Communications Organizations

ADC Kentrox
P.O. Box 10704
Portland, OR 97210
(503) 641-3341

Alcatel
1225 N. Alma
Richardson, TX 75081
(214) 996-5000

American Electronics Association
Japan Office
Yonbancho 11-4
Suite 101
Chiyoda-ku
Tokyo 102, Japan
(03) 3237-7195

American Mobile Satellite Corp.
1150 Connecticut Ave., NW
Washington, DC 20036
(202) 331-5858

American Personal
Communications
1025 Connecticut Ave., NW
Suite 904
Washington, DC 20036
(202) 296-0005

Ameritech Mobile
Communications
2000 West Ameritech Center Dr.
Hoffman Estates, IL 60195
(708) 234-9700

Andrew Corp.
10500 West 153rd St.
Orland Park, IL 60462
(708) 349-3300

AP Research
19672 Stevens Creek Blvd.
Suite 175
Cupertino, CA 95014
(408) 253-6567

Apple Computer
2025 Mariani Ave.
Cupertino, CA 95014
(408) 974-6790

Applied Engineering
P.O. Box 5100
Carrollton, TX 75011
(214) 241-0055

ARDIS
300 Knightsbridge Parkway
Lincolnshire, IL 60069
(708) 913-1215

Arianespace Inc.
700 13th St., NW
Suite 230
Washington, DC 20005
(202) 628-3936

Arthur D. Little, Inc.
Acorn Park
Cambridge, MA 02140
(617) 864-5770

AT&T
67 Whippany Rd.
Whippany, NJ 07981
(201) 386-7765

AT&T Consumer Electronics
5 Woodhollow Rd.
Parsippany, NJ 07054
(201) 581-3000

AT&T EasyLink Services
400 Interspace Parkway
Parsippany, NJ 07054
(201) 331-4000

AT&T Microelectronics
555 Union Blvd.
Allentown, PA 18103
(800) 372-2447

Audiovox Corp.
150 Marcus Blvd.
Hauppauge, NY 11788
(516) 231-7750

Bell Atlantic Mobile Systems
180 Washington Valley Rd.
Bedminster, NJ 07921
(908) 306-7583

Bell Communications Research
(Bellcore)
290 West Mt. Pleasant Ave.
Livingston, NJ 07039
(800) 523-2673

Bell Mobility
20 Carlson Court
Etobicoke, Ontario
Canada M9W 6V4
(416) 674-2220

BellSouth
3535 Colonnade Parkway
Birmingham, AL 35243
(205) 977-8281

BIS Strategic Decisions
One Longwater Circle
Norwell, MA 02061
(617) 982-9500

Blaupunkt
Robert Bosch Corp.
2800 S. 25th Ave.
Broadview, IL 60153
(708) 865-5200

BT (C.B.P.) Ltd.
Annandale House
1 Hanworth Rd.
Sunbury-on-Thames
Surrey, UK
44-932
76-5766

BT North America
2560 North First St.
P.O. Box 49019, MS-F25
San Jose, CA 95161
(800) 872-7654

Cable Television Laboratories, Inc.
1050 Walnut St.
Suite 500
Boulder, CO 80302

California Microwave, Inc.
Wireless Network Division
985 Almanor Ave.
Sunnyvale, CA 94086
(408) 732-4000

Calling Communications Corp.
1900 W. Garvey Ave. S.
Suite 200
West Covina, CA 91790
(818) 856-0671

Cantel Mobile Systems
40 Eglington Ave. East
Toronto, Ontario
Canada M4P 3A2
(416) 440-1300

Casio, Inc.
570 Mt. Pleasant Ave.
Dover, NJ 07801
(201) 361-5400

Cellular Data, Inc.
2860 West Bayshore Rd.
Palo Alto, CA 94303
(415) 856-9800

Cellular Foundation
1133 21st St., NW
3rd Floor
Washington, DC 20036
(202) 785-0081

Cellular Telecommunications Industry Association
1133 21st St., NW
3rd Floor
Washington, DC 20036
(202) 785-0081

Celsat, Inc.
532 S. Gertruda Ave.
Redondo Beach, CA 90277
(310) 316-6301

CenCall Communications
3231 S. Zuni
Englewood, CO 80110
(303) 761-4707

Chevalier (OA) Limited
2303-2305 Great Eagle Centre
23 Harbour Rd.
Wanchai
Hong Kong
(852) 827-2827

Clarion Corp. of America
661 W. Redondo Beach Blvd.
Gardena, CA 90247
(310) 327-9100

Communications Industry
Association of Japan
Sankei Bldg. Annex, 1-7-2,
Ohtemachi, Chiyoda-ku
Tokyo 100, Japan
(3) 3231-3156

COMSAT Mobile
Communications
22300 COMSAT Dr.
Clarksburg, MD 20871
(301) 428-4111

Constellation Communications,
Inc.
CIT Tower
Suite 200
2214 Rockhill Rd.
Herndon, VA 22070
(703) 733-2819

Cox Enterprises, Inc.
1400 Lake Hearn Dr.
Atlanta, GA 30319
(404) 843-5000

Creative Strategies Research
International
46 Old Ironsides Dr.
Suite 490
Santa Clara, CA 95054
(408) 748-3400

Cylink
310 North Mary Ave.
Sunnyvale, CA 94086
(408) 735-5817

Datacomm Research Co.
920 Harvard St.
Wilmette, IL 60091
(708) 256-1763

Electronic Industries Association
2001 Pennsylvania Ave., NW
Washington, DC 20006
(202) 457-4900

Electronic Messaging Association
1555 Wilson Blvd.
Suite 300
Arlington, VA 22209
(703) 875-8620

Ellipsat
1120 19th St., NW
Suite 480
Washington, DC 20036
(202) 466-4488

EO, Inc.
800A East Middlefield Rd.
Mountain View, CA 94043
(415) 903-8100

E-Plus Mobilfunk
Thyssen Trade Center
Hans-Guenther-Sohl-Strasse 1
4000 Duseldorf 1, Germany
(01149) 211-967-7590

Ericsson Business
Communications
1900 W. Crescent Ave.
Anaheim, CA 92801
(714) 533-5000

Ericsson GE Mobile
Communications, Inc.
15 E. Midland Ave.
Paramus, NJ 07652
(201) 265-6600

Ericsson Radio Systems, Inc.
740 East Campbell Rd.
Richardson, TX 75081
(214) 238-3222

Escort Division
Cincinnati Microwave, Inc.
5200 Fields-Ertel Rd.
Cincinnati, OH 45249
(513) 489-3900

European Community Delegation
2100 M St., NW
7th Floor
Washington, DC 20037
(202) 862-9500

Eurosat Distribution Ltd.
Head Office, London
1 Oxgate Centre, Oxgate Lane
Edgeware Rd., London NW2 7JG
44-081-452-6699

Eutelsat
Tour-Maine-Montparnasse 33,
Avenue du Maine, 75755
Paris, France
(011) 331 45 38 47 47

Ex Machina, Inc.
45 East 89th St.
#39-A
New York, NY 10128
(212) 831-3142

Federal Communications
Commission
1919 M St., NW
Washington, DC 20055
(202) 632-7557

The Freedonia Group
3570 Warrenville Center Rd.
Suite 201
Cleveland, OH 44122
(216) 921-6800

Frost & Sullivan
106 Fulton St.
New York, NY 10038
(212) 233-1080

Frost & Sullivan International
Sullivan House
4, Grosvenor Gardens
London SW1W ODH, UK
44-071-730-3438

Fujitsu Network Transmission
Systems
2801 Telecom Parkway
Richardson, TX 75082
(214) 690-6000

Fujitsu Personal Systems, Inc.
5200 Patrick Henry Dr.
Santa Clara, CA 95054
(408) 764-9489

GeoWorks
2150 Shattuck Ave.
Berkeley, CA 94704
(510) 644-0883

Glenayre Technologies, Inc.
4800 River Green Parkway
Duluth, GA 30136
(404) 623-4900

GO Corp.
919 East Hillsdale Blvd.
Suite 400
Foster City, CA 94404
(415) 345-7400

Goldstar
1850 W. Drake Dr.
Tempe, AZ 85283
(602) 752-2200

GRiD Systems Corp.
7 Village Circle
Westlake, TX 76262
(817) 491-5200

GTE Mobile Communications
245 Perimeter Center Parkway
P.O. Box 105194
Atlanta, GA 30348
(404) 391-8386

GTE PCS Group
600 N. Westshore Blvd.
Suite 600
Tampa, FL 33609
(813) 282-6154

GTE Spacenet
1700 Old Meadow Rd.
McLean, VA 22102
(703) 848-1391

Hayes Microcomputer Products, Inc.
5835 Peachtree Corners East
Norcross, GA 30092
(404) 840-9200

Herschel Shosteck Associates, Inc.
10 Post Office Rd.
Silver Spring, MD 20910
(301) 589-2259

Hewlett-Packard Co.
5301 Stevens Creek Blvd.
Building 51 Lower
Santa Clara, CA 95052
707-577-2663

Highway Master
16479 Dallas Parkway
Suite 300
Dallas, TX 75248
(214) 732-2500

Hong Kong Call Point
Century Plaza Three
19th Floor
Tai Koo Shing
Hong Kong
(852) 803-3663

Hong Kong Post Office
Telecommunications Branch
Sincere Bldg.
5th Floor
Central Hong Kong
(852) 852-9688

Hong Kong Telecom CSL Limited
City Plaza Phase Three
19th Floor
Hong Kong
(852) 803-8231

Hughes Network Systems
11717 Exploration Lane
Germantown, MD 20876
(301) 428-5500

Hutchison Paging Limited
Manlong House
611-615 Nathan Rd.
9th Floor
Kowloon
Hong Kong
(852) 710-6828

Hutchison Telecom
Great Eagle Centre
23 Harbour Rd.
27th Floor
Hong Kong
(852) 828-3230

Hutchison Telephone Limited
Citicorp Centre Ground Floor
18 Whitfield Rd.
North Point
Hong Kong
(852) 807-9765

IBM Personal Computer Co.
1000 N.W. 51st St.
Boca Raton, FL 33432
(407) 443-2000

Independent Telecommunications Network, Inc.
47th at Main
Gilbert Robinson Plaza Bldg.
4th Floor
Kansas City, MO 64112
(816) 561-9200

Industrial Computer Systems, Inc.
27972 Meadow Dr.
Evergreen, CO 80439
(303) 674-0700

In-Flight Phone Corp.
122 West 22nd St.
Suite 100
Oak Brook, IL 60521
(708) 573-2660

InfraLAN Technologies, Inc.
12 Craig Rd.
Acton, MA 01720
(508) 266-1500

INMARSAT
(International Maritime Satellite Organization)
Project 21 Office
40 Melton St.
London NM1 2EQ UK
44-71-387-9089

In-Stat Inc.
7418 East Helm Dr.
Scottsdale, AZ 85260
(602) 483-4440

Institute of Electrical and Electronics Engineers
445 Hoes Lane
Piscataway, NJ 08855
(908) 562-3000

Intel Corp.
2625 Walsh Ave.
Santa Clara, CA 95052
(408) 765-4483

Intelsat
3400 International Dr.
Washington, DC 20008
(202) 944-6963

InterDigital Communications Corp.
2200 Renaissance Blvd.
Suite 105
King of Prussia, PA 19406
(215) 278-7800

INTUG
(International Telecommunications Users Group)
18 Westminster Palace Gardens
London, W1
United Kingdom
(0441) 799-2446

Iridium, Inc.
1350 I St., NW
Suite 400
Washington, DC 20005
(202) 371-6880

Japan Electronic Industry
Development Association
3-5, Shibakeon, Minato-ku
Tokyo 105, Japan
(03) 3433-6296

Japan Radio Co.
Akasaka Twin Tower (Main)
17-22 Akasaka 2-chome
Minato-ku
Tokyo 107, Japan
(81) 3584-8836

Japan R&D Center for Radio
Systems
Bansui Bldg.
1-5-16 Toranoman
Minato-ku
Tokyo 105, Japan
(03) 3592-1101

Kenwood USA Corp.
2201 E. Dominquez St.
Long Beach, CA 90810
(310) 639-9000

Lexicus
345 Forest Ave.
Suite 45
Palo Alto, CA 94301
(415) 323-4771

Link Resources Corp.
79 Fifth Ave.
New York, NY 10003
(212) 627-1500

LOCATE
17 Battery Place
Suite 1200
New York, NY 10004
(212) 509-5595

Loral Aerospace Corp.
7375 Executive Place
Suite 101
Seabrook, MD 20706
(301) 805-0591

Loral Qualcomm Satellite
Services, Inc.
3825 Fabian Way
Palo Alto, CA 94303
(415) 852-4592

Lotus Development Corp.
Mobile Computing Division
One Rogers St.
Cambridge, MA 02142
(800) 448-2500

Marconi Communications Inc.
1800 Sunrise Valley Dr.
Reston, VA 22091
(703) 620-0333

Matra Marconi Space
7 rue Hermes
Parc Technologique du Canal
31520 Ramonville-St-Agne,
France
(33) 61-750565

McCaw Cellular
Communications, Inc.
P.O. Box 97060
Kirkland, WA 98083
(206) 827-4500

MCI Communications Corp.
1801 Pennsylvania Ave., NW
Washington, DC 20006
(202) 872-1600

Megahertz Corp.
4505 South Wasatch Blvd.
Salt Lake City, UT 84124
(800) LAPTOPS

Mercury Communications Ltd.
90 Long Acre
London WC2E 9NP UK
44-71-836-2449

Metricom, Inc.
980 University Ave.
Los Gatos, CA 95030
(408) 399-8200

Metriplex, Inc.
25 First St.
Cambridge, MA 02141
(617) 494-9393

Microsoft Corp.
One Microsoft Way
Redmond, WA 98052
(206) 936-4375

Minnesota Department of
Transportation
Research & Strategic Initiatives
117 University Ave.
Room 253
St. Paul, MN 55155
(612) 296-4935

Mitsubishi Electronics America,
Inc.
800 Biermann Court
Mt. Prospect, IL 60056
(708) 699-4317

Mitsubishi International Corp.
1500 Michael Dr.
Suite B
Wood Dale, IL 60191
(708) 860-4200

MobileComm
1800 E. County Line Rd.
Suite 300
Ridgeland, MS 39157
(601) 977-0888

Monicor Electronic Corp.
2964 N.W. 60th St.
Ft. Lauderdale, FL 33309
(305) 979-1907

Motorola, Inc.
Cellular Subscriber Group
600 North US Highway 45
Libertyville, IL 60048
(708) 523-5000

Motorola/EMBARC
Communication Services
1500 N.W. 22nd Ave.
Boynton Beach, FL 33426
(407) 364-2000

MTA/EMCI
1130 Connecticut Ave, NW
Suite 325
Washington, DC 20036
(202) 835-7800

Murata/Muratec
5560 Tennyson Pkwy.
Plano, TX 75024
(214) 403-3300

NASA
400 Maryland Ave., SW
Washington, DC 20546
(202) 358-1983

National Semiconductor
2900 Semiconductor Dr.
Santa Clara, CA 95054
(408) 721-5000

National Telecommunications & Information Administration
U.S. Department of Commerce
Washington, DC 20230
(202) 377-1866

NCR
WaveLAN Products
1700 South Patterson Blvd.
Dayton, OH 45479
(800) 225-5627

NEC America
Mobile Radio Division
383 Omni Dr.
Richardson, TX 75080
(800) 421-2141

The Netherlands Institute for Conformance Testing of Telecommunications Equipment (NKT)
P.O. Box 30605
2500 GP The Hague
070-3410582

NEXTEL Communications
201 Route 17 North
Rutherford, NJ 07070
(201) 438-1400

Nokia Mobile Phones
2300 Tall Pines Drive
Suite 120
Largo, FL 34641
(813) 536-4443

Northern Telecom
2221 Lakeside Blvd.
Richardson, TX 75208
(214) 684-8821

NovAtel
P.O. Box 1233
Fort Worth, TX 76101
(817) 847-2100

NYNEX Mobile Communications Co.
2000 Corporate Dr.
Orangeburg, NY 10962
(914) 365-7712

Oki Telecom
437 Old Peachtree Rd.
Suwanee, GA 30174
(404) 995-9800

Omnipoint Data Co.
2120 Hollowbrook Dr.
Colorado Springs, CO 80918
(719) 548-1200

O'Neill Communications, Inc.
Princeton Corporate Center
One Deerpark Dr.
Monmouth Junction, NJ 08852
(908) 329-4100

Orbital Communications
12500 Fairlakes Circle
Fairfax, VA 22033
(703) 818-3762

Orbitel Mobile Communications
The Keytech Centre
Ashwood Way
Basingstoke, Hamshire
RG23 8BG, UK
44-256
84-3468

Pacific Link Communications Limited
26 Harbour Rd.
China Resources Bldg.
30th Floor
Hong Kong
(852) 879-8644

PacTel Corp.
PacTel Corporate Plaza
2999 Oak Rd.
Walnut Creek, CA 94596
(510) 210-3645

Paging Network Inc.
4965 Preston Park Blvd.
Suite 600
Plano, TX 75093
(214) 985-4100

Palm Computing
4410 El Camino Real
Suite 108
Los Altos, CA 94022
(415) 949-9560

PanAmSat
1 Pickwick Plaza
Suite 270
Greenwich, CT 06830
(203) 622-6664

Panasonic Communications & Systems Co.
Two Panasonic Way
Secaucus, NJ 07094
(201) 348-7000

Paragraph International
1035 Pearl St.
Suite 104A
Boulder, CO 80302
(303) 443-8777

PCMCIA
1030 East Duane Ave.
Suite G
Sunnyvale, CA 94086
(408) 720-0107

PenMagic Software Inc.
310-260 West Esplanade
North Vancouver, BC
Canada V7M 3G7
(604) 988-9982

Performance Systems International, Inc.
510 Huntmar Park Dr.
Herndon, VA 22070
(703) 904-7187

Personal Communications Industry Association (Telocator)
1019 19th St., NW
Suite 1100
Washington, DC 20036
(202) 467-4770

Personal Communications Limited
64 Harcourt House
39 Glouchester Rd.
Wanchai
Hong Kong
(852) 860-8282

Personal Technology Research
296 Newton Ave.
Waltham, MA 02154
(617) 893-2600

Photonics Corp.
2940 N. First St.
San Jose, CA 95134
(408) 955-7930

Pioneer Electronics
2265 E. 220th St.
Long Beach, CA 90810
(310) 835-6177

Plexsys International Corp.
1245 Diehl Rd.
Naperville, IL 60563
(708) 505-0499

PowerTek Industries, Inc.
14550 East Fremont Ave.
Englewood, CO 80112
(303) 680-9400

Premier Telecom Products Inc.
600 Industrial Parkway
Industrial Airport, KS 66031
(913) 791-7000

Probe Research
3 Wing Dr., Ste. 240
Cedar Knolls, NJ 07927
(201) 285-1500

Proxim, Inc.
295 North Bernardo Ave.
Mountain View, CA 94043
(415) 960-1630

QUALCOMM, Inc.
10555 Sorrento Valley Rd.
San Diego, CA 92121
(619) 587-1121

Racotek, Inc.
7401 Metro Blvd.
Suite 500
Minneapolis, MN 55439
(612) 832-9800

RadioMail Corp.
Suite 275
2600 Campus
San Mateo, CA 94403
(415) 572-6000

RAM Mobile Data
10 Woodbridge Ctr. Dr.
Woodbridge, NJ 07095
(908) 602-5603

Raytheon Co.
528 Boston Post Rd.
Sudbury, MA 01776
(508) 440-2678

Rogers Cantel Mobile
Communications Inc.
Suite 2600
Commercial Union Tower
P.O. Box 249
Toronto Dominion Centre
Toronto, Ontario M5K 1J5
(416) 777-0880

Rose Communications Inc.
2390 Walsh Ave.
Santa Clara, CA 95051
(408) 727-7673

Samsung
3655 North First St.
San Jose, CA 95134
(800) 446-0262

Sanyo
21350 Lassen St.
Chatsworth, CA 91311
(818) 998-7322

Sharp Electronics Corp.
Sharp Plaza
Mahwah, NJ 07430
(201) 529-8200

Shintom West
20435 South Western Ave.
Torrance, CA 90501
(310) 328-7200

Sierra Wireless, Inc.
8999 Nelson Way
Burnaby, BC
Canada V5A 485
(604) 668-7328

Simware, Inc.
20 Colonnade Rd.
Ottawa, Ontario
Canada K2E 7M6
(613) 727-1779

SkyTel Corp.
1350 I St., NW
Suite 1100
Washington, DC 20005
(202) 408-7444

Slate Corp.
15035 North 73rd St.
Scottsdale, AZ 85260
(602) 443-7322

Snider Telecom
P.O. Box 4189
Little Rock, AR 72214
(501) 661-7600

Software Publishers Association
Pen Special Interest Group
1730 M St., NW
Suite 700
Washington, DC 20036
(202) 452-1600

Sony Corp. of America
One Sony Dr.
Park Ridge, NJ 07656
(201) 930-7066

Southwestern Bell Mobile
Systems
18111 Preston Rd.
Suite 900
Dallas, TX 75252
(214) 713-0000

SpectraLink Corp.
1650 38th St.
Suite 202E
Boulder, CO 80301
(303) 440-5330

Spectrum Ericsson
45 Crossways Park Dr.
Woodbury, NY 11797
(516) 822-9810

Sprint Cellular
O'Hare Plaza
8725 W. Higgins Rd
Suite 650
Chicago, IL 60631
(312) 399-2828

Stanford Telecommunications,
Inc.
2421 Mission College Blvd.
Santa Clara, CA 95056
(408) 748-1010

Strategies Unlimited
201 San Antonio Circle
Suite 205
Mountain View, CA 94040
(415) 941-3438

Sun Microsystems
2550 Garcia
Mountain View, CA 94043
(415) 960-1300

Symbol Technologies, Inc.
46 Wilbur Place
Bohemia, NY 11716
(800) SCAN-234

Tandy Corp.
700 One Tandy Center
Fort Worth, TX 76102
(817) 390-3300

Technologic Partners
419 Park Ave. South
Suite 500
New York, NY 10016
(212) 696-9330

TekNow, Inc.
4745 N. 7th St.
Suite 230
Phoenix, AZ 85014
(602) 266-7800

Telecommunications Industry
Association
2001 Pennsylvania Ave.
Suite 800
Washington, DC 20006
(202) 457-8737

Teledyne Microwave
1290 Terra Bella Ave.
Mountain View, CA 94043
(415) 960-8601

Telesystems SLW, Inc.
85 Scarsdale Rd.
Suite 201
Don Mills, Ontario
Canada M38 2R2
(416) 441-9966

Toshiba America, Inc.
9740 Irvine Blvd.
Irvine, CA 92713
(714) 583-3000

TRW
One Space Park
Redondo Beach, CA 90278
(310) 812-5227

U.S. Department of Commerce
Trade Statistics Division
Room 2217
14th St. and Constitution Ave.,
NW
Washington, DC 20230
(202) 377-4211

U.S. Paging Corp.
1680 Route 23 North
Wayne, NJ 07470
(201) 833-3952

United States Telephone
Association
900 19th St., NW
Suite 800
Washington, DC 20006
(202) 835-3100

United Telecommunications Inc.
2330 Shawnee Mission Parkway
Westwood, KS 66213
(913) 624-2641

Universal Paging Corp.
20 Broad Hollow Rd.
Melville, NY 11747
(516) 385-4100

US WEST NewVector Group
3350 161st Ave. S.E.
Bellevue, WA 98008
(206) 747-4900

VITA
1600 Wilson Blvd.
Suite 500
Arlington, VA 22209
(703) 276-1800

Westinghouse Electric Corp.
P.O. Box 17319
Baltimore, MD 21203
(410) 765-4146

Windata, Inc.
10 Bearfoot Rd.
Northboro, MA 01532
(508) 393-3330

Wireless Cable Association
International
2000 L St., NW
Suite 702
Washington, DC 20036
(202) 452-7823

WordPerfect Corp.
1555 North Technology Way
Orem, UT 84057
(801) 222-4050

Yankee Group
200 Portland St.
Cambridge, MA 02114
(617) 367-1000

Zenith Data Systems
2150 East Lake Cook Rd.
Buffalo Grove, IL 60089
(800) 553-0331

INDEX

O

P

Q

R

S

T

U

V

W

Z